CHEMISTRY OF METABOLISM AND ENERGETICS

CHEMISTRY OF METABOLISM AND ENERGETICS

Editor
N.P. SHARMA

KALPAZ PUBLICATIONS
DELHI-110052

CHEMISTRY OF METABOLISM AND ENERGETICS

(Chemistry)

ISBN: 81-7835-063-7

Published in 2001 in India by
Kalpaz Publications
C-30, Satyawati Nagar
Ashok Vihar Phase-III
Delhi-110052
E-Mail: kalpaz@hotmail.com

Laser Typesetting at : A.S. Printers
Printed at : Mehra Offset Printers, New Delhi

Contents

Chapter 1

Introduction

In the living systems, the energy is stored in the form of energy rich or high energy compounds which are mainly phosphates containing molecules. Whenever the energy is needed by metabolic process, these high energy compounds which are mainly phosphates containing molecules. Whenever the energy is needed by metabolic processes, these high energy compounds undergo decomposition and the energy liberated is utilised. Further, the syntheses of these compounds take place as result of various metabolic activities.

Suitable fuel is required to provide the energy that enables the animal to carry out its normal process. How the organism obtains this energy from its food is basic to the understanding of normal nutrition and metabolism.Death from starvation occurs when available energy reserves are depleted, and certain forms of malnutrition are associated with energy imbalance.

The rate of energy release, measured by the metabolic rate, is controlled by the thyroid hormones, whose malfunction is a cause of disease. Storage of surplus energy results in obesity, one of the most common diseases of occidental society. Energy of a system may be defined as the capacity of a system to do work. This capacity may be bound in the molecules. The bound form of energy is termed as potential energy.

Potential energy is generally measured' in terms of work that could be undertaken when the energy were released.

Potential energy may be of various types such as electrical, chemical, photic (light), atomic or positional. In the biological world, the main source of potential energy is the solar energy which gets fixed up in the form of chemical energy of carbohydrate molecules during photosynthesis in green plants.

Other organisms have to depend directly or indirectly upon this process (photo synthesis) for a constant supply of energy to maintain their structures and to perform their numerous functions. Energy is not only required for mechanical work, maintenance of body temperature and osmotic work but also to drive numerous synthetic reactions.

The molecules in motion possess capacity to carry out work. The energy possessed by a molecule or a body by virtue of its motion is termed as *kinetic energy*. This type of energy depends upon the mass of body and its velocity of motion in accordance to the following relation:

$$K.E = \frac{1}{2} mv^2$$

where K.E. represents the kinetic energy, in the mass of the body and v the velocity of the body.

Other forms of energy are as follows:

— Free energy.

— Internal energy.

The Joule is the unit of energy and is numerically equivalent to 0.239 calorie. One calorie is equal to 4.184 Joules (J) and one Kilocalorie is equal to 4.184 kilojoules (KJ).

Principles of thermodynamics

Biologic dystems conform to the general laws of thermodynamics. Many quantitative observations carried out on the interconversion of various forms of energy have helped scientists to the formulation of two fundamental laws of thermodynamics, the first and second. These laws have been found to be useful in understanding:

(a) the direction of a reaction, whether from left to right or vice versa,

(b) the accomplishment of work, whether useful or not, and

(c) whether the energy for driving a reaction must be delivered from an external source.

First Law of thermodynamics

The principle of conservation of energy which is based on cumulative experience of ages can be stated as:

> "Energy may be converted from one form to another, but it is impossible to create or destroy it".

Different attempts were made to disprove this principle by constructing a machine that may continue to operate without any expenditure of fuel and may thus create energy out of nothing (Perpetual motion machine of first kind). But all, attempts failed. The net result is that the principle of conservation of energy is universally accepted and is retitled as the first law of thermodynamics. There are various ways of enunciating the first law of thermodynamics. Here some of these are given as follows:

(i) *First statement:* It may be put as:

> "Energy of an isolated system must remain constant, although it may be transformed from one form to another.

As heat is given out to the surroundings in exothermic reactions, the internal energy of reacting system decreases while the energy of surroundings increases correspondingly. Therefore, the total energy of the system and the surroundings (i.e., isolated system) taken together remains unchanged. As heat is absorbed from the surroundings in exothermic reactions, the internal energy of the reacting system increases while the energy of the surroundings decreases correspondingly. Therefore, the total energy of the system and the surroundings taken together remains unchanged.

(ii) *Second statement:* It 'is as below:

> "Energy in one form, if it disappears, will make its appearance in an exactly equivalent quantity in another form.

For example, coal contains inherently a lot of internal energy. During burning this energy functions as chemical energy and yields heat. The beat is converted into the mechanical energy of the steam, which moves the engine to move a train. In this example, energy is disappearing in one form and appearing in another form in an exactly equivalent quantity.

Consider a system whose state on absorbing a quantity of beat 'q' changes from A to B. This heat is used in two ways:

(i) In increasing the internal energy of the system, i.e.

$$\Delta E = E_B - E_A$$

where E_A and E_B denote the internal energies associated with the system in its states A and B respectively, and

(ii) To do some external work 'w' by the system on its surroundings.

From the first law, we get

$$\text{Heat absorbed by the system} = \text{Increase in its internal energy} + \text{Work done by the system}$$

$$\therefore \qquad q = \Delta E + w$$

$$\Delta E = q - w \qquad (1)$$

Differential form of the first law. For an infinitesimal process, Eq. (1) may become as follows:

$$dE = \delta q + \delta w \qquad (2)$$

Here an infinitesimal process may be defined as a process involving only infinitesimal changes in thermodynamic variable defining the system. In Eq. *(2)* the differential is distinguished in notations by *d* from δ. This can be understood by the following discussion.

The change in internal energy is dependent on path by which it is accomplished and the previous history of the system. Thus, change in internal energy is an exact differential. So in mathematical terms, an exact differential is always denoted by a notation *d, e.g., dE* in equation (2).

The *'q'* and *'w'* are not state functions because changes in their magnitude for a process depend upon:

(i) the path by which the change is accomplished, and

(ii) the previous history of the system.

The first law of thermodynamics establishes an exact relation between heat and work. It tells us that a certain quantity of heat will produce a definite amount of work and vice-versa. The first law of thermodynamics denies that work or energy can be produced from nothing. Work cannot appear without disappearing of heat. The discovery of Albert Einstein that mass is convertible into energy, however, makes it essential to make a change in the definition of the first law of thermodynamics. It is possible to convert mass into energy in accordance to the following equation:

$$E = mc^2$$

where E is the energy produced by the destruction of mass m and c is the velocity of light. The modified first law of thermodynamics may be stated as follows:

> "The total mass and energy of an isolated system remains unchanged."

As matter to energy or energy to matter conversions are of little importance in the biological systems, it means that the first law of thermodynamics has no valid applications in biochemistry.

There are three main limitations of first law of thermodynamics, These are as follows :

a) The first law of thermodynamics simply establishes equivalence between different form of energy. However, this law does not tell us under what conditions and to what extent it is possible to bring about conversion of one form of energy into the other.

   This can he understood from the following examples: i) We know that heat is flowing from a hot to cold body. First law simply tells us that the quantity of heat lost by the hot body is equal to the heat gained by the cold body. It never tells us that heat can only flow from the hot to the cold body and not in the reverse direction. ii) Zinc dissolves spontaneously in sulphuric acid, a definite amount of energy being liberated. The reverse process is not spontaneous. Without external influence, the reaction proceeds in only one direction. From the above examples we can predict that there must be some laws besides the first law which should from the basis of predicting the direction of a certain reaction to occur, and, if so, and to what extent. This, however, comes within the scope of the second law of thermodynamics.

b) The first law of thermodynamics is a qualitative statement which does not prelude the existence of either a heat engine or a refrigerator. The first law does not contradict the existence of a 100% efficient heat engine ora self-acting refrigerator. In practice, these two are not attainable. These phenomena are recognised and this led to the formulation of a law governing these devices. It is called the second law of thermodynamics.

c) The first law fails to explain why chemical reactions do not proceed to completion. For example, consider the reaction $CO(g)+H_2O(g)$ $CO_2(g)+H_2(g)$. If originally one mole of CO and one mole of H_2O were in the reaction vessel, then if the reaction went to completion, one mole of CO_2 and one mole H_2 would be produced. If the reaction vessel is examined at equilibrium, the yield is less than 100%. However, this comes within the scope of second law of thermodynamics.

It is the energy associated with a system by virtue of its molecular constitution and the motion of its molecules. The contribution of energy *due* to molecular constitution is know it as *internal potential energy* and the contribution of energy due to the motion of molecules is known as *internal kinetic* energy. Hence the total internal energy of a system is given by the sum of two types of energies. i.e.,

$$\frac{\text{Total internal}}{\text{energy}} = \frac{\text{Internal potential}}{\text{energy}} + \frac{\text{Internal kinetic}}{\text{energy}}$$

A fixed quantity of any substance, under a given set of conditions, has got a definite amount of internal energy associated with it. This amount is different for different substance. It is, therefore, regarded as *an extensive property*. Internal energy is symbolized by E or U.

Second law of thermodynamics

All the natural processes tend to have a tendency to go equilibrium. The second law of thermodynamics is related to the equilibrium state of a process. Second law of the thermodynamics is generalisation of certain experiences about heat engines and refrigerators, It has been stated in a number of ways. but all the, statements are logically equivalent to one another.

Kelvin-Planck statement: "It is impossible to obtain a continuous supply of work from a body or engine that can transfer heat from a single heat reservoir". The above statement is a negative statement. By this statement, it means that a single reservoir at a single temperature cannot continuously transfer heat into work. It means that there should be two reservoirs for any heat engine. One reservoir (called the source) is taken at a higher temperature while the other reservoir (called the sink) is taken at a lower temperature.

According to this statement, zero degree absolute temperature is not attainable as no heat is rejected to the sink at zero degree Kelvin. Suppose an engine works between any temperature higher than zero, degree Kelvin. It implies that it uses a single reservoir which controdicts the Kelvin-Planck's statement of the second law. Similarly, it can be said that no engine can be 100% efficient.

Clausius statement. Heat flows from hot to cold body due to difference of temperature. However, in a refrigerator, heat gets absorbed from a body at lower temperature and gets rejected to a body at higher temperature. Here heat is being flown through some outer agency, i.e., it is not feasible to cause heat to flow from a colder to a hotter body without absorbing from some external source. This led Clausius to state the second law of the thermodynamic in the following words : "It is impossible to transfer for a self acting machine, unaided by any external agency to convert heat continuously, from one body at lower temperature to a body at a higher temperature."

Planck statement: "It is impossible to construct a device that will work in a complete cycle and convert heat into work without making any change in the surroundings.

Thomson statement: In a heat engine, the working substance takes heat from hot body; a part of it is converted into work while remaining is given to the sink. No engine has so far been constructed which continuously takes heat from a single body and changes the whole of it to work without making any change in the system. Hence the presence of a cold body is a necessity for converting heat into work. This led Thomson to state that "it is impossible to obtain a continuous supply of work by cooling a body to a temperature lower than that of coldest of its surroundings." Diffusion of a gas always takes place from a region of higher to one at a lower pressure, until the pressure becomes uniform.

The second law of thermodynamic may be stated in various ways. All the statements, appear to be different from one another. But a closer look at them, reveals that they are all the modified forms of the same fundamental idea that, all forms of energy are convertible into heat but the heat so obtained cannot be converted into other forms of energy by any process," The difference between the first and second laws of thermodynamics lies in the fact that *the first law* is concerned with the accounting of the various kinds of energy involved in a given process, while the second law is concerned with the availability of the energy of a given system for doing useful work.

The second law of thermodynamics introduces the concept of entropy which may be regarded as the amount of energy unavailable for conversion into work. The term entropy was produced by Clausius and is denoted by symbol S. In simple words, entropy of a system may be regarded as the amount of energy unavailable for conversion into work.

It is, however, very difficult to define actual entropy of a system precisely. Therefore, more generally the change in the entropy of a system is defined rather than the absolute entropy of the system. The increase in entropy dS in the course of an infinitesimal change is equal to δq_{rev}, the heat absorbed when the change is carried out in a reversible manner divided by the absolute temperature T, i.e.

$$dS = \frac{\delta q_{rev}}{T} \tag{3}$$

Although δq_{rev} is not an exact differential, it has a definite value for a reversible, isothermal change, and dS as defined by equation (3) is an exact differential. The entropy S of a system is thus a state function. If heat change takes place at different temperatures, than we have

$$dS = \frac{\delta Q_1}{T_1} + \frac{\delta Q_2}{T_2} + \frac{\delta Q_3}{T_3} + \ldots = \Sigma \frac{\delta Q_{rev}}{T} \tag{4}$$

Eq. (3) may be written as follows

$$dS = \frac{\delta Q_{rev}}{T}$$

or $$S = \frac{\delta Q_{rev}}{T} \tag{5}$$

From equation (5) it means that "the entropy difference between two states of a given system is equal to the summation of all the Q_{rev} /T terms of all the steps which are necessary to bring about a reversible transformation between these two states, where q_{rev}, denotes the heat absorbed reversibly and T the corresponding temperature.

For a finite change at constant *temperature dS becomes* ΔS, δq_{rev} becomes q_{rev} and Eq. (1) becomes as follows:

$$S = \frac{q_{rev}}{T}$$

$$S = \frac{\text{Heat change}}{\text{Absolute temperature}}$$

$$= \frac{J}{K} = JK^{-1}$$

ΔS is an extensive property and its value, therefore, depends upon the amount of substance involved. Therefore, the unit of entropy will be $JK^{-}{}_{1}$ $mol^{-}{}_{1}$ (S.I. units). In both physical and biological world, the entropy is always changing. Whenever there occurs spontaneous change in process the entropy increases.

A characteristic of a spontaneous process (such as the flow of heart in metal base from the hot to the cold end, diffusion of one gas into another, the dissolution of a solid by a solvent etc.) is that it is accompanied by an increase in the 'disorder' or randomness' of the molecules of entropy.Hence entropy may be regard as" a measure of disorder or randomness of the molecular arrangement in a system.

A spontaneous process always proceeds from a less probable state to a more probable state by the interplay of natural forces. We know that a spontaneous process is always accompanied by an increase in the entropy, we can therefore identify the entropy as a function of the probability of the state. Since we know that both the entropy, and the thermodynamic probability increase simultaneeusly in a process we can therefore consider the state of equilibrium as the state of maximum probability.

Free energy

In biological systems, the free energy is the most important type of energy. It is denoted by G. It may be regarded as a measure of the potential energy of the substance but it is a quantity that cannot be measured directly. However, its change accompanying chemical reaction or a process can he measured. It is denoted by ΔG. The change in free energy (ΔG is that portion of the total energy change in a system which is available for doing work., i.e., it is the useful work.) The change in free energy ΔG of a reaction is calculated by taking the difference between the sum of the free energies of the products and that of reactants.

$$A + B \rightleftharpoons C + D \tag{6}$$

$$\Delta G = \text{Free energy} = (G_c+G_D) - (G_A=G_B) \tag{7}$$

When the free energy change in a reaction is negative (ΔG negative), the reaction is called exergonic and when the free energy change is positive, (ΔG positive), the reaction is called endergonic. An example of exergonic precess is the process of respiration while an example of endergonic process is the process or photosynthesis. It is often said that the reactions with a negative ΔG can occur spontaneously but the sign and magnitude of ΔG give no information on the rate of a reaction. Thus, for the complete oxidation of a glucose according to the equation.

$$C_6H_{12}O_6 + 6O_2 \longrightarrow 6CO_2 + 6H_2O$$

ΔG = 2870 kJ per mole of glucose; in spite of this large negative ΔG, glucose is quite stable in air and the rate spontaneous reaction being negligibly small. Further if ΔG is zero, a state of chemical equilibrium exists and no net chemical change takes place.

Combination of First Second Law of Thermodynamics

As the entropy charges of chemical reactions could not be readily measurable, the entropy is not regarded as criterion whether a biochemical process can take place spontaneously or not. Furthermore, for spontaneity, both entropy changes should be 'known. These difficulties could be overcome by using a different thermodynamic function, called the free energy, which is denoted by the symbol G. (In older literature, however, the symbol F is used in its place). The symbol G has been given in honour Josiah Gibbss, an American physical chemist. who single-handedly created a large portion of chemical thermodynamics In 1878, Gibbss introduced the free energy function by combining the firs', and second laws of thermodynamics in the form of following equation:

$$\Delta G = \Delta H - T\Delta S \quad (8)$$

where, ΔG = the change in free energy of a creative system

ΔH = the change in heat content or enthalpy of this system

T = the absolute temperature at which the process is taking place, and

ΔH = the change in entropy of this system.

This equation gives quantitative interrelationship between the changes in energy, heat and entropy in chemical reactions taking place at constant temperature (T) and pressure (P), the conditions prevailing in biological systems. The term $T\Delta S$ is that fraction of AH which cannot be converted into useful work. The ΔG reveals the free energy change or the theoretically available useful work.

Naturally, in most cases the system has been 'inefficient' and not all of the theoretically available work can be utilised. However, the properties of the surroundings do not take part into this process: The enthalpy change, ΔH is given by the following equation:

$$\Delta H = \Delta E + p\Delta V \qquad (9)$$

where, ΔE = the change in internal energy of a reaction, and

ΔV = the change in volume of this reaction.

As the volume changes, ΔV has been found to be small for nearly all, biochemical reactions, hence ΔH has been nearly equal to the change in internal energy, ΔE. Therefore, Equation (9) becomes as follows:

$$\Delta G \approx AE - T\Delta S \qquad (10)$$

Thus, the change in free energy of a reaction, ΔG will depend both on the change in internal energy and on the change in entropy of the system. The ΔG has been a valuable criterion in determining whether a reaction can take place spontaneously.

(i) If ΔG if negative, the reaction takes place spontaneously with loss of free energy, i.e., it is exergonic. If, in addition ΔG is of great magnitude, the reaction goes virtually to completion and has been essentially irreversible.

(ii) If, however, ΔG is positive, the reaction proceeds only if free energy can be gained, i.e., it is endergonic. If, in addition, ΔG is of high magnitude, the system is stable having little or no tendency for a reaction to take place.

(iii) If ΔG is zero, the system is at equilibrium and no net change takes place. It means that the system is stable with little or no tendency for a reaction to occur.

With regard to the free energy change ΔG of a reacting system two more points have to be understood.

1. Firstly, the ΔG of a reaction is dependent only on the free energy of the products minus that of the reactants. The ΔG of a reaction is independent of the path of transformation. Obviously, it means that the mechanism of a reaction has no effect on ΔG. As an instance, the value of ΔG has been the same for the oxidation of glucose to CO_2 and H_20 whether it occurs by combustion or by a series of enzyme-catalysed reactions.

2. Secondly, the value of ΔG gives no information about the rate of a reaction. A negative ΔG reveals that a reaction can take place spontaneously, but it does not signify that it will take place at perceptible rate. As we already know that, the rate of a reaction rather depends on the free energy of activation (ΔG = ≠, which is unrelated to ΔG.

In order to get a feeling for the magnitude of changes in various forms of energy, an actual example of aerobic oxidation of glucose may be taken into consideration. Pie living cells carry out oxidation of glucose in the presence of oxygen to CO_2 and H_2O at constant temperature and pressure.

$$C_6H_{12}O_6+6O_2 \longrightarrow 6CO_2+6H_2O$$

If we assume that the temperature is 25°C (or 298 K) and the pressure is 1.0 aim. (or 760 mm Hg), which are standard conditions in thermodynamic calculations, the following energy changes occur per molecule of glucose oxidised—

$\Delta G = -686{,}000$ cal/mol (i.e., the free energy of the reacting molecules has decreased)

$\Delta H = -673{,}000$ cal/mol (i.e., the reacting molecules have released heat)

Eq. (10) may also be written as:

$$\Delta S = \frac{\Delta H - \Delta G}{T} \qquad (11)$$

On substituting the above values in Eq. (11), we get

$$\Delta S = \frac{-673{,}000 - (-686{,}000)}{298}$$

$$= 44 \text{ cal/deg}$$ i.e., the entropy of the universe has increased

$$= 184.096 \text{j/deg}$$

Standard free energy change and equilibrium constant

In a model reaction

$$A + B \rightleftharpoons C + D \qquad (12)$$

the free energy change, ΔG of this reaction will be given by

$$\Delta G = \Delta G^\circ + RT \log_e \frac{[C]\,[D]}{[A]\,[B]} \qquad (13)$$

where, $\Delta G°$ = Standard free energy change

R = Gas constant

T = Absolute temperature

[A], [B], [C] and [D] = Molar concentrations (i.e., activities) of the reactants.

The standard free energy change ($\Delta G°$) of a chemical reaction may be defined as the free energy change when the reactions and products are in their standard states (298 K, I atm pressure). Thus, the ΔG of a reaction depends on the nature of the reactants (expressed in $\Delta G°$ term) and on their concentrations (expressed in logarithmic terms), as shown in Eq. (13).

At equilibrium, $\Delta G = 0$. Equation then becomes as follows:

$$\Delta G = \Delta G° + RT \log_e \frac{[C]\,[D]}{[A]\,[B]} \quad (14)$$

$$\Delta G° + RT \log_e \frac{[C]\,[D]}{[A]\,[B]} \quad (15)$$

The equilibrium constant under standard conditions, K'_{eq} for the reaction $A + B \rightleftharpoons C + D$, will be given as follows:

$$K'_{eq} \quad \frac{[C]_{eq}\,[D]_{eq}}{[A]\,[B]}$$

At Equilibrium, there is no change in free energy (ΔG= 0); therefore, Eq. (4) becomes as follows:

$$0 = DG° + RT \ln \frac{[C]_{eq}\,[D]_{eq}}{[A]_{eq}\,[B]_{eq}}$$

By changing logarithms to the base 10, the above equation becomes as follows:

$$0 = \Delta G^\circ + 2.303 \text{ RT} \log K'_{eq} \quad (16)$$

$$\Delta G^\circ = -2.303 \text{ RT} \log K'_{eq} \quad (17)$$

Eq. (17) reveals that if it is possible to determine equilibrium constant for any reaction, the standard free energy change can also be calculated. Conversely if ΔG° is known, the actual free energy ΔG for any reaction can also be calculated. Eq. (17) can be put in the following form:

$$K'_{eq} = 10 \quad (18)$$

Substituting R = 1.98×10^{-3} kcal mol^{-1} $degree^{-1}$ and T = 298° K (corresponding to 25°C) gives

$$K'_{eq} = 10 \quad (19)$$

when $\Delta^\circ G$ has been expressed in kcal/mol. Hence the standard free energy, ΔG° and the equilibrium constant, K'_{eq} are related by a simple expression. For example, a change in equilibrium constant by a factor of 10 causes a change in standard free energy of 5.69 kj/mol at 25°C. At 37°C, however, the change in standard free energy would be of 5.69 kj/mol. Values of ΔG° may be expressed in joules or calories per mole.

If ΔG° is negative (i.e., the products have less free energy than the reactants), the reaction will take place to form the products under standard conditions, as all chemical reactions will tend to go in that direction causing a decrease in the free energy of the system. If ΔG° is positive (i.e., the products of the reaction contain more free energy than the reactants), the will proceed in the reverse direction if we start with 1M concentration of all components. To be more explicit, reactions having a negative ΔG° take place forward in the direction written when they start with all reactants and products at 1.0 M, until they reach equilibrium

Table 1. Numerical relationship between equilibrium constants of chemical reactions and their standard free energy changes at 25°C

K'_{eq}	ΔG° *(kcal/mol)*	ΔG° *(kj/mol)*
10^{-5} or 0.00001	+6.82	+28.53
10^{-4} or 0.0001	+5.46	+22.84
10^{-3} or 0.001	+4.09	+17.12
10^{-2} or 0.01	+2.73	+11.42
10^{-1} or 0.01	+1.36	+5.69
1 .	1	0
10^{1} or 10	–1.36	–5.69
10^{2} or 100	–2.73	–11.42
10^{3} or 1,000	–4.09	–17.12
10^{4} or 10,000	–5.46	–22.84
10^{5} or 1,00,000	–6.82	–28.53

Table 2. Relationship among K'_{eq} ΔG° and the directions chemical reactions understand conditions

When K'_{eq} is	ΔG *is*	*Starting 1M components, the reactions*
>1.0	Negative	Proceeds forward
1.0	Zero	Remainsat equilibrium
<1.0	Positive	Proceeds backward

Reactions with a positive ΔG° will take place in the reverse of the of the direction written when they start with all components at 1.0 M. Table 2 these relationships. Two more points should be understood:

1. Biochemical reactions occur near pH 7.0. Hence, PH 7.0 is conventionally designated the standard pH in biochemical energetics. The standard free energy change at pH 7.0 in biochemical energetics has been designated by the symbol, $\Delta G°$.

2. The basic SI (Syste'me International) unit of energy is joule (J). However, in biology and medicine, beat and energy measurements are equally expressed in calories (cal). The interrelationship between calories and joules is 1.000 cal = 4.184 J.

Biochemist's standard state

In physical chemistry, the standard state refers to the situation in which all the reactants and products are at unit molar (or molal) concentration. In biochemistry, we follow the same procedure except that we define the hydrogen-ion concentration for the standard state to lie 10^{-7} M because the physiological pH is about 7. Consequently, the change in the standard Gibbs free energy according to these two conventions will be different for reactions involving uptake or liberation of hydrogen ions. We shall therefore replace $\Delta G°$ with $\Delta G°$ in discussing biochemical process. Consider the reaction

$$A + B \longrightarrow C + xH^+$$

The Gibbs free energy change (ΔG) for the process is given by

$$\Delta G = \Delta G° + RT\ In \frac{\frac{[C]}{1M} \quad \frac{[H^+]}{1M}}{\frac{[A]}{1M} \quad \frac{[B]}{IM}}$$

where 1M represents the physical chemists' standard state of solute in solution. Since the biochemists standards state for H^+ ions is 10^{-7} *M*, the Gibbs free energy change for the same process is now given by

$$\Delta G = DG^\circ + RT \ In \frac{\frac{[C]}{1M} \quad \frac{[H^+]}{10^{-7} M}}{\frac{[A]}{1M} \quad \frac{[B]}{IM}}$$

Note that regardless of the convention for the standard state, ΔG must remain unchanged. From the last two equations we obtain This means that for reactions producing H^+ ions, ΔG° is grater than ΔG° by 40.0 kj per mole of H^+ ions released.Hence the reaction is more spontaneous at pH 7 than at pH 0. On the other hand, if H^+ ion appears as the reactant,

$$C + xH^+ \longrightarrow A + B$$

we can show that

$$\Delta G^\circ = \Delta G^\circ - 40.0 \text{ KJ}$$

and this reaction will be more spontaneous at pH 0 than at pH 7. For reactions not involving H^+ ions, ΔG° is equal to ΔG°.

Standard free energy changes at pH 7.0 or ΔG°

In order to calculate ΔG° we will consider an example of isomerization of dihydroxy-acetone phosphate (DHAP) to glyceraldehyde 3-phophate (G-3-P). It has been one of the reaction of glycolysis. At equilibrium, the ratic of glyceraldehyde 3-phosphate to dihydroxy- acetone phosphate has been found to be 0.0475 at 250C (298 0 K) and pH 7.0. Hence

$$K'_{eq} = 0.0475$$

The standard free energy change for this reaction is, then, calculated by using Eq. (20) as follows:

$$\Delta G^0_{.} = -2.303\ RT\ \log_{10} K'_{eq} \qquad (20)$$

$$= -\ 2.303 \times 1.98 \times 10^{-3} \times 298 \times \log_{10} (0.0475)$$

$$= +\ 1.8\ \text{kcal/mol}$$

If the initial concentration of ΔHAP has been 2×10^{-4} M and the initial concentration of G-3-P is 3×10^{-6} M′ then ΔG can be calculated by using Eq. (21) as follows:

$$\Delta G = \Delta G^{\circ} + RT \log_e \frac{[C]\ [D]}{[A]\ [B]}$$

$$= 1.8\ \text{kcal/mol} + 2.303\ RT \log_{10} \frac{3 \times 10^{-6}\ M}{2 \times 10^{-4}\ M}$$

$$= 1.8\ \text{kcal/mol} - 2.5\ \text{kcal/mol}$$

$$= -0.7\ \text{kcal/mol}$$

$$= -2.93\ \text{kJ/mol}$$

The negative value for ΔG reveals that isomerization of dihydroxyacetone phosphate to glyceraldehyde 3-phosphate can take place spontaneously, when these compounds are present at concentrations mentioned above. Although ΔG has been negative for this reaction, yet the value of ΔG° has been positive. It is important to note that the magnitude of ΔG for reaction (whether smaller, larger or the same as ΔG°) has been found to depend on the concentrations of the reactants. The criterion of spontaneity for a reaction is ΔG and not ΔG°. Table 3 lists the standard free energy changes for some chemical reactions.

Table 3. Standard free energy changes for representative chemical reactions under standard conditions (that is, at 25°C and pH. 7.0)

Reaction type Reactions	ΔG^{o} *(in kcal/mol)*	ΔG^{o} *(in kj/mol)*
Elimination of Water		
Malate → Fumarate+ H_2O	+ 0.75	+3.138
Rearrangement		
Glucose 1- phosphate → Glucose 6- phosphate	−1.74	−7.280
Fructose 6-phosphate Glucose 6- phosphate	−0.40	−1.6736
Hydrolysis		
Esters		
Enthy1 acetate +H_2O → Ethanol+ Acetate	−4.7	−19.665
Glucose 6- phosphate+ H_2O Glucose→+Phosphate	−3.3	−13.807
Amides and peptides		
Glutamine + H_2O → Glutamate + NH_4^+	−3.4	−14.225
Glycyglycine +H_2O → 2 Glycine	−2.2	−9.205
Glycosides		
Maltose +H_2O → 2 Glucose	−3.7	−15.481
Lactose H_2O → Glucose+ Galactose	−3.8	−15.899
Acid anhydrides		
Acetic anhydride +H_2O → 2 Acetate	−21.8	−91.211
ATP+ H_2O → ADP+ Phosphate	−7.3	−30.543
Oxidations with molecular oxygen		
Glucose + $6O_2$ → $6CO_2$+ $6H_2O$	−686	−2870.224
Palmitic acid + $23O_2$ → $16CO_2$ +16 H_2O	−2338	−9782.192

Hydrolysis of esters, amides, peptides and glycosides, as well as rearrangements, and elimination takes place with relatively small standard free energy change.

Hydrolysis oı acid anhydrides takes place with relatively large decreases in standard free energy, whereas oxidation of organic compounds to CO_2 and H_2O occurs with rather huge decreases in standard free energy.

ΔG and ΔG°

It is better if we understand the difference between the free energy change, ΔG and the standard free energy change, ΔG^0 of a chemical reaction. The actual free energy change, ΔG of a chemical reaction has been a function of the conditions of concentration, pH and temperature under which the reaction is occurring.

Moreover the ΔG of an ongoing chemical process always becomes negative, becomes smaller (i.e., less negative) as the reaction proceeds and becomes zero at the point of equilibrium, revealing that no more work can be done by the reaction. The value of ΔG decreases with time as the reaction proceeds because the actual concentrations of the reactants will be getting smaller and those of the resultants getting larger.

In every spontaneous chemical or physical process, the free energy of the reacting system always decreases, i.e., ΔG is negative. On the other hand., the value of standard free energy change. ΔG°, for a chemical reaction has been characteristic and unchanging, and may be positive, negative or zero, depending on the equilibrium constant of the reaction, ΔG° is, thus an immutable constant and predicts in which direction and how far a given reaction will go in order to reach equilibrium when it takes place under standard condition, i.e., when the initial concentration of all components is 1.0 M, the pH is 7.0 and the temperature is 25°C.

The overall free energy change for a series of reactions has, been equal to the sum of the free energy changes of the individual steps. Let us consider the two consecutive reactions where each reaction is having its own equilibrium constant and a characteristic standard free energy change ΔG°_1 and ΔG°_2.

$$A \longrightarrow B \qquad \Delta G_1^\circ = -8 \text{ kcal/mol}$$

$$B \longrightarrow C \qquad \Delta G_2^\circ = +5 \text{ kcal/mol}$$

As the reactions are sequential, the intermediate product B cancels out and the with its standard free energy change, AG°s may be put as follows:

$$A \longrightarrow C \qquad \Delta G^\circ_s = \Delta G^\circ_1 + G_2^\circ$$
$$= -8 + (+5)$$
$$= -3 \text{ kcal/mol}$$
$$= -12.552 \text{ kj/mol}$$

Under standard conditions, A can get spontaneously converted into B as G is negative. However, the conversion of B into C, under standard conditions has been not thermodynamically feasible. But as the free energy changes are additive, the conversion of A into C has a ΔG° value of -12.552 kj/mol, which obviously implies that A can be converted into C spontaneously under standard conditions.

Thus, the above two sequential reactions get coupled by the intermediate product. B. In other words, a thermodynamically unfavourable reaction can be driven by a thermodynamically favourable reaction. Two sequential steps from glycogen break down in muscles will demonstrate this fact more clearly.

$$\text{Glucose 1- phosphate} \xrightarrow{\text{Phospho-glucomutase}} \text{Glucose 6-phosphate} \qquad \Delta G^\circ_1 = -1.74 \text{ kcal/mol}$$

$$\text{Glucose 6- phosphate} \xrightarrow{\text{Glucose phospate isomerase}} \text{Fructose 6-phosphate} \qquad \Delta G^\circ_1 = +0.40 \text{ kcal/mol}$$

On adding the two reactions, we obtain

Glucose 1- phosphate ⟶ Fructose 6-phosphate

This has a standard free energy change value,

$$\Delta G^\circ_s = \Delta G^\circ_1 + \Delta G_2^\circ$$
$$= -1.74 + (+\ 0.40)$$
$$= -1.36 \text{ kcal/mol}$$
$$= -5.6902 \text{ kj/mol}$$

As $\Delta G^\circ s'$ is negative, glucose 1 phosphate gets converted into fructose 6-phosphate in the muscles.

Coupled reactions

In biochemical reactions, two or more reactions may be taking place together in such a way that an exergonic reaction may be used to drive an endergonic one, Such reactions are known as coupled reactions. The simplest type of coupling may be represented by the equation:

$$A + C \longrightarrow B + D + \text{Heat}$$

The conversion of metabolite A to metabolite B takes place with the release of energy. It is coupled to another reaction, wherein energy is required to convert metabolite C to metabolite D. Many instances of this sort of coupling are seen in metabolism when the energy derived from an exergonic reaction (usually an oxidation) is used to drive an endergonic synthetic reaction. This sort of coupling requires both reactants to contain a common obligatory intermediate.

$$A + C \longrightarrow I \longrightarrow B + D$$

This type of coupling should have a built-in mechanism for biological control of the rate at which oxidative, processes are allowed to take place because the existence of a common obligatory intermediate for both the exergonic and endergonic reactions allows the rate of utilisation of the product of the synthetic path (D) 'to determine by mass action the rate of which A is oxidised. This type of coupling is provided by dehydrogenation reactions which are coupled to hydrogenation by an intermediate carrier.

An alternative method of coupling an exergonic to an endergonic process may involve the synthesis of a new compound of high-energy potential in the exergonic reaction and then this new compound may he incorporated into endergonic reaction thus effecting a transference of free energy from the exergonic to the endergonic pathway.

In order to maintain living processes, all organisms must get free energy from their environment. The autotrophic, organisms get supply of free energy by coupling their metabolism to some simple exergonic process in their surroundings. For example, green plants utilise the energy of sunlight. On the other hand, heterotrophic organisms get free energy by coupling their metabolism to the breakdown of complex organic molecules in their environment. The special carrier of free energy has been adenosine triphosphate (ATP), ATP plays a central role in the transference of free energy from the exergonic (energy yielding) to the endergonic (= energy requiring) processes in the cells.

During breakdown of energy rich foodstuffs or fuel molecules, some of the free energy gets harnessed to make ATP from adenosine diphosphate (ADP) and inorganic phosphate (P_1) a process that requires input of free energy. ATP then donates much of its chemical energy to energy needing processes (biosynetheses transport etc.) by undergoing a breakdown to ADP and P_1. A large number of coupled reactions involve ATP or another nucleoside triphosphate groups. In its reactions in the cell, it functions as the Mg^{+2} complex. The importance of phosphates in the intermediary metabolisms was realised by Fritz in 1941 and he introduced the concept of "high-energy phosphates" and "high-energy phosphate bond".

High-energy phosphates

ATP is frequently described as an energy-rich compound with bounds called high -energy bounds. A high energy bond is one that liberates a large amount of energy when hydrolyzed (>5 kcal/mole). Two such high energy groups exist in the ATP anhydride linkage. Let's use a block diagram of ATP so that we can focus our attention on its high-energy groups. The anyhydrides can be hydrolzed. Upon hydrolysis of 1 mole of ATP to adenosine diphosphate (ADP), about 7300 cal (30, 500 J) are liberated under the operating conditions of the cell.

By exothermically hydrolyzing ATP to ADP or AMP, a cell liberates energy which can be used to power endothermic processes. The hydrolysis of ADP to AMP is exothermic but does not occur frequently in cellular metabolism. In order to indicate the presence of the high energy phosphate group, Lipman employed the symbol NP to indicate high energy phosphate bond. This symbol is indicative of the fact that the group attached to the bond, when transferred to an appropriate acceptor, results in transfer of the larger of quantity energy. Due to this reasons tire group transfer potential is used to indicate "high-energy bond".

For Instance, ATP has two high energy phosphate groups and ADP has one while the phosphate bond in AMP (adenosine monophosphate) is of the low-energy type because it is a normal ester link. Cells also contain other high-energy compounds in addition to ATP. When hydrolyzed, these compounds also liberate more than 5 kcal/mole.

Structural formulas for several of these compounds are given below. Phosphocreating (creatine phosphate) is an important storage compound. The phosphate group is added directly and reversibly to creative from ATP under the influence of the enzyme creative kinase. These reversible reactions are particularly important in skeletal muscle tissue.

During muscular work, ATP is needed for muscle construction. When the muscle is at rest, ATP concentration is high and the forward reaction above is favoured, a phosphate group is transferred to certain, forming phosphocreatine. In this way, creatine phosphate.

During muscular work, ATP concentration decreases as ATP is converted to ADP. The increased ADP concentration causes the reverse reaction above to be favoured: direct phosphorylation of ADP by phosphocreatine occurs. The hydrolysis of 1 mole of phosphocreatine liberates 10,300 cal (43, 100)

$$\text{Phosphocreatine} + \text{water} \longrightarrow \text{creatine} + \text{phosphate} + 10.300 \text{ cal}$$

Unlike the acid anhydride linkage hydrolysed in ATP the P–N bond hydrolysed in phosphocreatine is an amide linkage. The standard free energy of hydrolysis of a number of biochemically important phosphates has been given in Table 4. From Table 4 it can be seen that the value for the hydrolysis of the terminal phosphate of the ATP of 36.8 kJ (8.8) (kcal) per mol (as it is also for the terminal phosphate of ATP) divides the list into 2 groups.

(i) *Low energy compounds:* This group has $\Delta G^{\circ\prime}$ values which are smaller than that of ATP. This group is exemplified by the ester phosphates found in the intermediates of glycolysis.

Table 4. Standard free energy of hydrolysis of some organophosphates of biochemical importance

		$\Delta G°$	
	Compound	*kJmol⁻¹*	*kcal/mole*
(i)	Phosphoėnol pyruvate	–61.1	–14.8
(ii)	Carbamoy 1 Phosphate	–51.4	–12.3
(iii)	1,3-Biphosphogly cerate (to 3 phosphogly cerate)	–49.3	–11.8
(iv)	Creatine phosphate	–13.1	–10.3
	ATP ⟶ ADP + Pi	–36.8	–8.8
(v)	Glucose 1-phosphate	–20.9	–5.0
(vi)	Fluctose 6-phosphate	–15.9	–3.8
(vii)	Glucose 6-phosphate	–13.8	–3.2
(viii)	Glycerol 3- phospate	–9.2	–2.2

(ii) *High energy compounds:* The group has $\Delta G°'$ values which are higher than that of ATP. This group is exemplified by anhydrides (e.g., ATP, ADT, the 1-phosphate of 1, 3-biphosphogy-cerate), enolphosphates (e.g., phosphoenolpyruvate) and phosphogunidines (e.g., creative phosphate, arginine phosphate). Other biologically important compounds that are classified as "high-energy compounds" are thiol esters involving coenzyme. A (e.g., acetyl CoA), acyl carrier ' protein, amino acid esters involved in protein synthesis, Sadenosy Imethionine (active methionine) and UDPG (uridine diphosphate glucose).

Due to its middle position in the Table 4, ATP acts as a donor of high-energy phosphate to those compounds which are below it in the Table 4. Similarly, ADP can accept high-energy phosphate to form ATP from those compounds which are above ATP in the Table 4. In reality, an ATP/ADP cycle connects these processes which generate ~ P to those processes which utilise Δ9. In Table 4, there is another group of compounds which act as storage forms of high-energy phosphate. These are creatine phosphate (phosphagen) and arginine phosphate.

In biological systems ATP serves as a source of energy for many of the endergonic reactions. Also, ATP get synthesised in many exergonic reactions. When an energonic reaction gets coupled with an exergonic reaction through ATP, the overall effect of this coupling involves the transfer of kinetic energy from one system to another. Several enzymes are known which facilitate these coupling reactions.

The coupling of reactions so as to bring about efficient transfer of energy is quite common in biological systems. This is illustrated by considering a reaction which is the first in the glycolysis pathway.

For example, the formation of glucose 6 phosphate from glucose arid phosphoric acid requires energy with $\Delta G^\circ + 13.8$ kJ mol^{-1}

(i) Glucose + Pi → Glucose 6-P + H_2O; $\Delta G^{\circ\prime} = + 13.8$ kJ mol^{-1}

The above reaction would not proceed as such under physiologic condition. In order to take place, the reaction must be coupled with another reaction which should be more exergonic than the phosphorylation of glucose which is endergonic, Such a reaction is nothing but the hydrolysis of the terminal phosphate of ATP.

(ii) ATP $\longrightarrow$ ADP + Pi ; $\Delta G^{\circ\prime}$ = 36.8 kJ mol^{-1}

If (i) and (ii) get coupled in a reaction catalysed by hexokinase, there occurs ready phosphorylation of glucose in a highly exergonic reaction that under physiologic conditions is far from equilibrium and thus irreversible for practical purposes.

$$\text{Glucose + ATP} \xrightarrow{\text{Hexokinase}} \text{Glucose 6.P + ADP}$$
$$\Delta G^{\circ} = +\ 23.2\ \text{kJ mol}^{-1}$$

Many "activation reactions" have been, found to follow this pattern. By using energy coupling a complex compound could be apparently which violates the second law of thermodynamics. But the algebraic sum of energy of all the components of coupled reaction is negative, i.e., he sum of energy liberated has been found Lo be more than the sum of energy consumed. In biological systems, the hydrolysis of energy rich compounds is generally accompanied by an endergonic reaction 0 that the energy liberated in the former reaction is used fully without getting wasted. This type of control needs numerous enzyme.; which are present in the cell. The high energy compounds get dissociated in the presence of enzymes.

Interconversion of adenine nucleotides

The enzyme adenylate kinase is present in most cells. It catalyzes the Interconversion of ATP and AMP on the one hand and ADP on the other:

$$\text{ATP + AMP} \xleftrightarrow{\text{Adenylate Kinase}} \text{2ADP}$$

This reaction has 3 functions:

(1) It allows high-energy phosphate in ADP to be used in the synthesis of ATP.

(2) It allows AMP, formed as a consequence of several activating reactions involving ATP, to be recovered by rephosphorylation to ADP.

(3) It allows AMP to increase in concentration when ATP becomes depleted and act as a metabolic (allosteric) signal to increase the rate of catabolic reactions, which in turn leads to the generation of more ATP.

When ATP Reacts to Form AMP, Inorganic Pyrophosphate (PPi) is Formed. This occurs, for example, in the activation of long-chain fatty acids.

$$ATP + CoA{\cdot}SH + R{\cdot}COOH \xrightarrow[\text{Synthetase}]{\text{Acyl-CoA}} AMP + PP_1 + R{\cdot}CO - SCoA$$

This reaction is accompanied by loss of free energy as beat, which ensures that the activation reaction will go to the right; this is further aided by the hydrolytic splitting of PP_1, 1 catalysed by inorganic pyrophosphatase, a reaction that itself has a large $\Delta G°$ OF – 27.6 kJ/mol. Note that activations via the pyrophosphate pathway result in the loss of 2 P rather than one P as occurs when ADP and P_1 are formed.

ATP synthesis

Energy is released upon ATP hydrolysis. This means that the reverse process is endothermic and energy is stored in ATP during its formation.

$$\frac{7300 \text{ cal} + ADP\ P_1}{8000 \text{ cal} + AMP + PP_1} \qquad \frac{ATP + HOH}{ATP + HOH}$$

Where does the energy come from that's used in ATP formation? It arises during catabolism in which cells chemically break down large metabolite molecule into simpler substances of lower energy. A net release of energy occurs and some of the energy is stored by conversion of AMP and ADP to ATP.

The formation of ATP by capturing energy liberated during catabolism is called oxidative phosphorylation. The majority of ATP is produced by this process. Oxidative phosphorylation occurs through the action of the respiratory chain, also known as the chain or the electron transport system (ETS). The components of the ETS are found on the inner membrane surface of mitochondria. The enzymes of the ETS function as a sequence in which hydrogens are passed along carriers. Ultimately the hydrogens combine with oxygen to produce water.

Hydrogen atoms can be imagined to consist of component hydrogen ions and electrons.

$$H + H = 2e + 2H^+$$

The loss of hydrogen with its associated electron from a substance is called oxidation; gain of hydrogen and, its electron is called reduction. The formation of water involves the reaction

$$2H^+ + 2e^- + \frac{1}{2} O_2 \longrightarrow H_2O$$

A biological requirement exists for hydrogen ion and electron carriers. Two kinds of carriers are present in the electron transport system: (1) coezymes, which are electron and hydrogen ion carriers, and (2) cytochromes', which are exclusively electron carriers. The coenzymes presently known to participate in the ETS are (1) nicotinamide adenine dinucleotides. (NAD and NADP), (2) flavin nucleotides (FAD and FMN), and (3) coenzyme Q (CoQ).

In order to maintain living processes, all organisms must obtain supplies of free energy from their environment. *Autotrophic* organisms couple their metabolism to some simple exergonic process in their surroundings, e.g., green plants utilise the energy of sunlight, and some autotrophic bacteria utilise the reaction $Fe^{2+} \longrightarrow Fe^{3+}$. On the other hand. heterotrophic organisms obtain free energy by coupling their metabolism to the breakdown of complex organic molecules in their environment. In all of these processes, ATP plays a central role in the transference of free energy from the exergonic to the endergonic processes.

The importance of phosphates in intermediary metabolism became evident with the discovery of the chemical details of glycolysis and of the role of ATP, adenosine diphosphate (ADP and inorganic phosphate ($P_{1'}$) in this process ATP was considered a means of transferring phosphate radicals in the process of phosphorylation.

The role of ATP in biochemical energetics was.indicated in experiments demonstrating that ATP and creatine phosphate were broken down during muscular contraction and that their resynethesis depended, on supplying energy from oxidative processes in the muscle. It was not until Lipmann introduced the concept of "high-energy phosphates" and the "high-energy phosphate bond that the role of these compounds in bioenergetics was clearly appreciated. So far we have presented a rather general and qualitative description of how some of the concepts can be applied to help us understand the nature of biochemical reactions

Under what conditions is thermodynamics applicable to biology? The answer to this question is not a trivial one, and at present there is much debate on this subject. To begin with, we note that biochemical reactions are usually discussed in terms of standard free energies (ΔG°). Strictly, the direction of a reaction such as

$$A + B \rightleftharpoons C + D$$

is given by ΔG, where

$$\Delta G \; \Delta G^{O} = + RT \; In \; \frac{[C]\,[D]}{[A]\,[B]}$$

Only in the case where the reactants and products are in their standard states can we employ ΔG^{O}. Thus the fact that the hydrolysis of ATP at 298 K and pH 7 results in a decrease of standard free energy equal to 30.5 kJ mol^{-1} does not necessarily mean that this is also the value for ΔG. In living cells the physiological temperature, pH, concentration of reactants and products, as well as metal ions can vary from system to system, and these factors must affect the value of ΔG.

It is difficult and in some cases impossible to accurately measure the concentration of various, species in solution. Nevertheless, reasonable estimates can be made of the concentrations in many cases and hence of the sign of ΔG. Another point that has often beep raised is that thermodynamics deals only with, closed systems at equilibrium. Living systems are open systems and they are maintained at steady state rather than at equilibrium. In fact, any cell at equilibrium is dead cell.

The treatment of open systems requires irreversible thermodynamics, which is beyond the scope of this book. Reactions in a living cell, like any other reactions, can be classified into two categories those that are thermodynamically controlled and those that are kinetically controlled. As an example of the former, let us consider the synthesis of a dipeptide from two amino acids:

alanine + glycine ⟶ alanylgycine ΔG^{O} = 17.2 kJ

The corresponding equilibrium constant for this reaction is about 1 × 103 at 298 K. Clearly, such a process will not proceed by an appreciable extent if left on its own.

However, if this process could be coupled to the hydrolysis of ATP by the action of an enzyme, the reaction would proceed from left to right. Such a reaction is said to be thermodynamically controlled because the reaction itself is not spontaneous, and the energy(or free energy) of reaction must be supplied from an outside source. A kinetically controlled reaction is one in which the overall ΔG is negative (and hence thermodynamically favourable), but the rate is negligibly small in the absence of appropriate enzyme catalysis. The phosphorylation of glucose to glucose 6-phosphate (coupled to ATP hydrolysis) is cerainly an exergonic process, but the reaction occurs at a very slow rate in the absence of the enzyme hexokinase.

Spontaneity of a process is the free energy,

Spontaneous does not mean "fast"; some spontaneous processes can take a long time to occur. Before this, we used the term "energetically favourable" to indicate spontaneous processes. The laws of thermodynamics can be used to predict whether any change involving transformations of energy will take place. An example of such a change is a chemical reaction in which covalent bonds are broken and new ones formed. Another example is the formation of noncovalent interactions, such as hydrogen bonds or hydrophobic interactions, when proteins fold to produce their characteristic three-dimensional structures.

The tendency of polar and nonpolar substances to exist in separate phases is a reflection of the energies of interaction between the individual molecules—in other words, the thermodynamics of the interaction. The most useful criterion for predicting the spontaneity of a process is the free energy, which has the symbol G. (Strictly speaking, the use of this criterion requires conditions of constant temperature and pressure, which are usual in biochemical thermodynamics.)

It is not possible to measure absolute values of energy, only the changes in energy that occur during a process. The value of the change in free energy (ΔG, where the symbol Δ indicates change) gives the needed information about the spontaneity of the process under consideration. The free energy of a system decreases in a spontaneous (energy-releasing) process, so ΔG is negative ($\Delta G < 0$). Such a process is called exergonic, meaning that energy is released.

When the free energy change is positive ($\Delta G > 0$), the process is nonspontaneous (the reverse of such a process is spontaneous). For a nonspontaneous process to occur, energy must be supplied. Nonspontaneous processes are also called endergonic, meaning that energy is absorbed. For a process at equilibrium, with no net change in either direction, the free energy change is zero ($\Delta G = 0$). The sign of the free energy change, ΔG, indicates the direction of the reaction.

$\Delta G < 0$ Spontaneous exergonic—energy released

$\Delta G = 0$ Equilibrium

$\Delta G > 0$ Nonspontaneous endergonic—energy required

An example of a spontaneous process is the aerobic metabolism of glucose, in which glucose reacts with oxygen to produce carbon dioxide, water, and energy for the organism.

$$\text{Glucose} + 6O_2 \rightarrow 6CO_2 + 6H_2O \quad \Delta G < 0$$

An example of a nonspontaneous process is the phosphorylation of ADP (adenosine diphosphate) to give ATP (adenosine triphosphate). This reaction takes place in living organisms because metabolic processes supply energy.

The choice of standard conditions is arbitrary. For a process under standard conditions, all substances involved in the reaction are in their standard states, in which case they are also said to be at unit activity. For pure solids and pure liquids, the standard state is the pure substance itself For gases the standard state is usually taken as a pressure of 1 atmosphere of that gas. For solutes, the standard state is usually taken as 1 molar concentration. Strictly speaking, these definitions for gases and for solutes are approximations, but they are valid for all but the most exacting work.

For any general reaction

$$aA + bB \rightarrow cC + dD$$

we can write an equation that relates the free energy change (ΔG) for the reaction under any conditions to the free energy change under standard conditions (ΔG°); the superscript °refers to standard conditions. This equation is

$$\Delta G = \Delta G^{0} + RT \ln \frac{[C]^{c}[D]^{d}}{[A]^{a}[B]^{b}}$$

In this equations, the square brackets indicate molar concentrations, R is the gas constant (8.31 J mol^{1} K^{-1}), and T is the absolute temperature. The notation In refers to natural logarithms (to the base "e") rather than logarithms to the base 10, for which the notation is log. This equation holds under all circumstances; the reaction does not have to be at equilibrium.

The value of ΔG under a given set of conditions depends on the value of ΔG° and on the concentration of reactants and products (given by the second term in the equation). In most biochemical reactions it is better to speak of ΔG°, which is independent of concentrations. There is only one ΔG° for a reaction at a given temperature.

When the reaction is at equilibrium, ΔG = 0, and thus

$$0G = \Delta G^0 + RT\ln\frac{[C]^c[D]^d}{[A]^a[B]^b}$$

$$\Delta G^0 = RT\ln\frac{[C]^c[D]^d}{[A]^a[B]^b}$$

The concentrations are now equilibrium concentrations, and this equation can be rewritten

$$\Delta G^{\circ} = -\ RT\ \ln\ K_{eq}$$

where K_{eq} is the equilibrium constant for the reaction. We have here a relationship between the equilibrium concentrations of reactants and products and the standard free energy change. Once we have determined the equilibrium concentrations of reactants by any convenient method, we can calculate the equilibrium constant, K_{eq}. We can then calculate the standard free energy change, ΔG°, from the equilibrium constant.

Biochemical applications

The calculation of standard free energy changes includes the stipulation that all substances be in standard states, which for solutes can be approximated as a concentration of 1 molar. If the hydrogen ion concentration of a solution is 1 molar, the pH is zero.

The interior of a living cell is, in many respects, an aqueous solution of the cellular components, and the pH of such a system is normally in the neutral range. Biochemical reactions in the laboratory are usually carried out in buffers that are also at or near neutral pH. For this reason it is convenient to define for biochemical practice a modified standard state, one that differs from the original standard state only by the change in hydrogen ion concentration from 1 M to 1 × 1^{-7} M, implying a pH of 7.

Wren free energy changes are calculated on the basis of this modified standard state, they are designated by the symbol ΔG°. Let us assume that the relative concentrations of reactants have been determined for a reaction carried out at pH 7 and 25°C (298 K). Such concentrations can be used to calculate an equilibrium constant, K'_{eq}, which in turn can be used to determine the standard free energy change, ΔG°, for the reaction.

A typical reaction to which this kind of calculation can be applied is the hydrolysis of ATP at pH7, yielding ADP, monohydrogen phosphate ion (written as P_i), and H^+:

$$ATP + H_2O \rightarrow ADP + P_i + H^+$$

$$K'_{eq} = \frac{[ADP][P_I][H^+]}{[ATP]} \quad pH7,\ 25°C$$

The concentrations of the solutes are used to approximate their activities, and the activity of the water is one. An experiment is done in the laboratory to determine K_{eq} for this reaction; the value obtained is 2.23×10^5. Substituting $R = 8.31$ J mol^{-1} K^{-1} $T = 298$ K, and In $K_{eq} = 12.32$,

$$\Delta G^{\circ\prime} = -RT \ln K'_{eq}$$

$$\Delta G^{\circ\prime} = -(8.31 \text{ J mol}^{-1} \text{ K-1}) (298 \text{ K}) (12.32)$$

$$\Delta G^{\circ\prime} = -3.0500 \times 104 \text{ J mol}^{-1} = -30.5 \text{ kJ mol}^{-1} = -7.29 \text{ kcal mol}^{-1}$$

In addition to illustrating the usefulness of a modified standard state for biochemical work, the negative value of $\Delta G^{\circ\prime}$ indicates that the reaction of hydrolysis of ATP to ADP is a spontaneous process in which energy is released.

Hydrophobic interactions

Hydrophobic interactions have important consequences in biochemistry. Large arrays of molecules can take on definite structures as a result of hydrophobic interactions. A phospholipid bilayer is one such array. We have seen that phospholipids are molecules that have polar head groups and long nonpolar tails of hydrocarbon chains.

Under suitable conditions, a double-layer arrangement is formed so that the polar head groups of many molecules face the aqueous environment, while the nonpolar tails are in contact with each other and are kept away from the aqueous environment. These bilayers form three-dimensional structures called liposomes. Such structures are useful model systems for biological membranes, which consist of similar bilayers with proteins embedded in them. The very existence of membranes depends on hydrophobic interactions.

Hydrophobic interactions are a major factor in the folding of proteins into the specific three-dimensional structures required for their functioning as enzymes, oxygen carriers, or structural elements. The order of amino acids automatically determines the three-dimensional structure of the protein. It is known experimentally that proteins tend to be folded so that the nonpolar hydrophobic side chains are sequestered from water in the interior of the protein, whereas the polar hydrophilic side chains lie on the exterior of the molecule and are accessible to the aqueous environment. What makes hydrophobic interactions favourable.

Hydrophobic interactions are spontaneous processes. The entropy of the universe increases when hydrophobic interactions occur.

$$\Delta S_{universe} > 0$$

As an example, let us assume that we have tried to mix the liquid hydrocarbon hexane (C_6H_{14}) with water and we have obtained not a solution but a two-layer system, one layer of hexane and one of water. Formation of a mixed solution is nonspontaneous, and the formation of two layers is spontaneous. Unfavourable entropy terms enter into the picture if solution formation requires the creation of ordered arrays of solvent, in this case water.

The greater degree of hydrogen bonding of the solvent water molecules around the nonpolar solute molecules introduces a higher degree of order, more like the lattice of ice than liquid water, and thus a lower entropy. The required entropy decrease is too large for the process to take place. Therefore, nonpolar substances do not dissolve in water; rather, nonpolar molecules associate with one another by hydrophobic interactions and are excluded from water.

Role of coenzymes

Coenzymes are nonprotein substances that take part in enzymatic reactions and are regenerated for further reaction. They include metal ions and a number of organic compounds; both groups of coenzymes can take part in oxidation-reduction reactions. Oxidation—reduction reactions are discussed at length in textbooks of general and inorganic chemistry, but the oxidation of nutrients by living organisms to provide energy requires its own special treatment. The description of redox reactions in terms of oxidation numbers, which is widely used with inorganic compounds, can be used to deal with the oxidation of carbon-containing molecules. However, our discussion will be more pictorial and easier to follow if we write equations for the half reactions, then concentrate on the functional groups of the reactants and products and on the number of electrons transferred. An example is the oxidation half reaction for the conversion of ethanol to acetaldehyde.

Writing the Lewis electron-dot structures for the functional groups involved in the reaction helps us keep track of the electrons being transferred. In the oxidation of ethanol, there are 12 electrons in the part of the ethanol molecule involved in the reaction and 10 electrons in the corresponding part of the acetaldehyde molecule; two electrons are transferred to an electron acceptor This type of "bookkeeping" is useful for dealing with biochemical reactions.

Many biological oxidation reactions, like this example, are accompanied by the transfer of a proton (H^+). The oxidation half reaction has been written as a reversible reaction because the occurrence of oxidation or reduction depends on the other reagents present. Another example of an oxidation half reaction is that for the conversion of NADH, the reduced form of nicotinamide adenine dinucleotide, to the oxidized form, NAD^+. This substance is an important coenzyme in many reactions.

The structure of NADH; the nicotinamide portion, the functional group involved in the reaction, is indicated in red. Nicotinamide is a derivative of nicotinic acid, one of the B-complex vitamins. A similar compound is NADPH (for which the oxidized form is $NADP^+$). To simplify writing the equation for the oxidation of NADH, only the nicotinamide ring is shown explicitly, with the rest of the molecule designated as *R*. The two electrons that are lost when NADH is converted to NAD^+ can be considered to come from the bond between carbon and the lost hydrogen, with the nitrogen lone-pair electrons becoming involved in a bond. Note that the loss of a hydrogen and two electrons can be considered as the loss of a hydride ion (H:-) by NADH and is sometimes written that way. The equations for both the reaction of NADH to NAD^+ and that of ethanol to acetaldehyde have been written as oxidation half reactions. If ethanol and NADH were mixed in a test tube, no reaction could take place because there would be no electron acceptor.

If, however, NADH were mixed with acetaldehyde, which is an oxidized species, a transfer of electrons could take place, producing ethanol and

NAD^+. The NADH is oxidized while the acetaldehyde is reduced. Another important electron acceptor is FAD (flavin adenine dinucleotide), which is the oxidized form of $FADH_2$. The symbol $FADH_2$ explicitly recognizes that protons (hydrogen ions) as well as electrons are accepted by FAD. Several other coenzymes contain the flavin group; they are derived from the vitamin riboflavin (vitamin B_2).

Oxidation of nutrients to provide energy for an organism cannot take place without reduction of some electron acceptor (i.e., an oxidizing agent). The ultimate electron acceptor in aerobic oxidation is oxygen; we shall encounter intermediate electron acceptors as we discuss metabolic processes. Reduction of metabolites plays a significant role in living organisms in anabolic processes. Important biomolecules are synthesized in organisms by many reactions in which a metabolite is reduced while the reduced form of a coenzyme is oxidized.

Homeostasis

In living cells, very few individual 'reactions ever reach equilibrium because these reactions are often coupled with each other. This places a continuous stress preventing equilibrium from being reached. You would be dead if all cells in your body achieved equilibrium. Consider that your body temperature serves as a useful guide to your physiological condition. The normal 98.6°F value and its relative constancy results from the net metabolic changes in your body even though your physical activities are not constant. You cat, sleep, work, walk, sit, sing—all different activities—and somehow your body accommodates these diverse functions without a drastic change in body temperature. The constant conditions which are maintained In the might be termed equilibria. That word, however, has come to have fairly exact meaning as applied to relatively simple physicochemical states, in closed systems where known forces are balanced. The coordinated physiological which maintain most of the steady states in the organism are so complex, and so peculiar to living beings involving, as they may, the brain and nerves, that heart lung, kidneys and spleen, all working cooperatively that we have suggested special designation for these states: homeostasis. The word does not imply something set immobile, stagnation. It means a condition a condition may vary, but which is relatively constant.

Chapter 2

Electron Transport in Metabolism

The electron transport chain is actually four closely related enzyme complexes embedded in the inner mitochondrial membrane. In a series of oxidation-reduction transfers, they conduct electrons along the membrane from one complex to another until the electrons reach their final destination where they combine with molecular oxygen to reduce O_2 to $2H_2O$. The energy of electron transport can then be used by these same enzyme complexes to pump protons across the inner membrane out into the intermembrane space.

The reverse flow of protons back through the membrane into the inner matrix can be used to generate ATP Also embedded in the inner membrane is an ATP synthase complex that binds ADP and phosphate to synthesize ATP The flow of protons through the ATP synthase from the intermembrane space to the inner matrix releases the new ATP that has been synthesized.

This process is very similar to the production of ATP by photosynthesis in the thylakoid membrane of the chloroplast in green plants. Aerobic metabolism is a highly efficient way for an organism to extract energy from nutrients. In eukaryotic cells the aerobic processes (including conversion of pyruvate to acetyl-CoA, the citric acid cycle, and electron transport) all occur in the mitochondria, whereas the anaerobic process, glycolysis, takes place outside the mitochondria in the cytosol.

Production of ATP in the mitochondria is the result of oxidative phosphorylation, in which ADP is phosphorylated to give ATP. The production of ATP by oxidative phosphorylation (an endergonic process) is separate from electron transport to oxygen (an exergonic process), but the reactions of the electron transport chain are strongly linked to one another and are tightly coupled to synthesis of ATP by phosphorylation of ADP.

Chemiosmotic coupling is the name given to this mechanism. Oxidative phosphorylation gives rise to most of the ATP production associated with the complete oxidation of glucose. The NADH and $FADH_2$ are oxidized to NAD^+ and FAD, which can be used again in various metabolic pathways. Oxygen, the ultimate electron acceptor, is reduced to water; this completes the process by which glucose is completely oxidized to carbon dioxide and water.

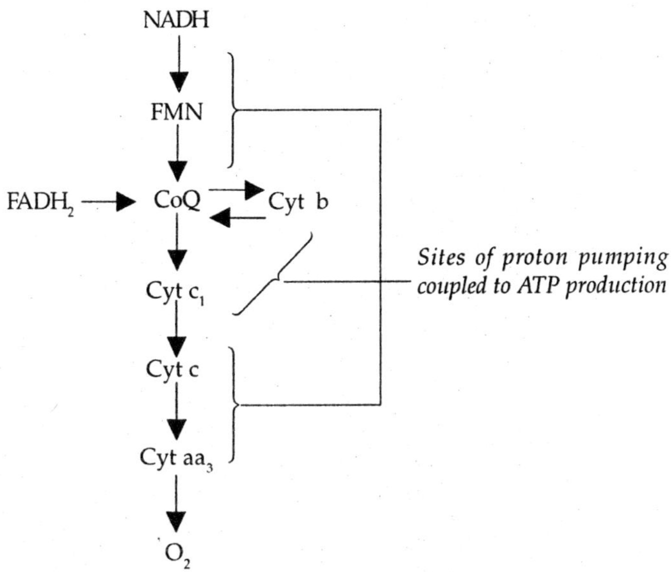

Fig. 1. schematic representation of electron transport chain showing complete series of oxidation-reduction reactions

The complete series of oxidation-reduction reactions of the electron transport chain is presented in schematic form in Figure 1. A particularly noteworthy point about electron transport is that on average 2.5 moles of ATP are generated for each mole of NADH that enters the electron transport chain, and on average 1.5 moles of ATP are produced for each mole of $FADH_2$. The general outline of the process is that NADH passes electrons to coenzyme Q as does $EADH_2$, providing an alternative mode of entry into the electron transport chain. Electrons are then passed from coenzyme Q to a series of proteins called cytochromes, designated by lowercase letters, and eventually to oxygen.

Electron transport form NADH to O_2

Intact mitochondria isolated from cells can carry out all the reactions of the electron transport chain; the electron transport apparatus can also be resolved into its component parts by a process called fractionation. Four separate respiratory complexes can be isolated from the inner mitochondrial membrane.

These complexes are multienzyme systems; in the last chapter we encountered other examples of such multienzyme complexes, such as the pyruvate dehydrogenase complex and the α–ketoglutarate dehydrogenase complex. Each of the respiratory complexes can carry out the reactions of a portion of the electron transport chain.

Complex I: The first complex, NADH-CoQ oxidoreductase, catalyzes the first steps of electron transport, namely the transfer of electrons from NADH to coenzyme Q (CoQ). This complex is an integral part of the inner mitochondrial membrane and includes, among other subunits, several proteins that contain an iron-sulphur cluster and the flavoprotein that oxidizes

NADH. The flavoprotein has a flavin coenzyme, FMN, which differs from FAD in not having an adenine nucleotide. The reaction occurs in several steps with successive oxidation and reduction of the flavoprotein and the iron-sulphur moiety. The first step is the transfer of electrons from NADH to the flavin portion of the flavoprotein:

$$NADH + H^+ + E\text{—}FMN \rightarrow NAD^+ + E\text{-}FMNH_2$$

where the notation E-FMN indicates that the flavin is covalently bonded to the enzyme. In the second step the reduced flavoprotein is reoxidized and coenzyme Q is reduced to $CoQH_2$ as a pair of electrons is passed on .

Coenzyme Q is also called ubiquinone. The electrons are passed first to the iron-sulphur clusters and then to coenzyme Q.

$$E\text{-}FMNH_2 + 2\ Fe\text{—}S_{oxidized} \rightarrow E\text{-}FMN + 2\ Fe\text{—}S_{reduced} + 2H+$$

$$Fe\text{—}S_{reduced} + CoQ + 2H^+ \rightarrow Fe\text{—}S_{oxidized} + COQH_2$$

The notation Fe-S indicates the iron-sulphur clusters. The overall equation for the reaction is

$$NADH + H^+ + CoQ \rightarrow NAD+ + CoQH_2$$

This reaction is one of the three responsible for the proton pumping that creates the pH (proton) gradient. The standard free energy change indicates that the reaction is strongly exergonic, releasing enough energy to drive the phosphorylation of ADP to ATP.

Coenzyme Q itself is mobile; that is to say, it is free to move in the membrane and to pass the electrons it has gained to the third complex for further transport to oxygen. We shall now see that the second complex also transfers electrons from an oxidizable substrate to CoQ.

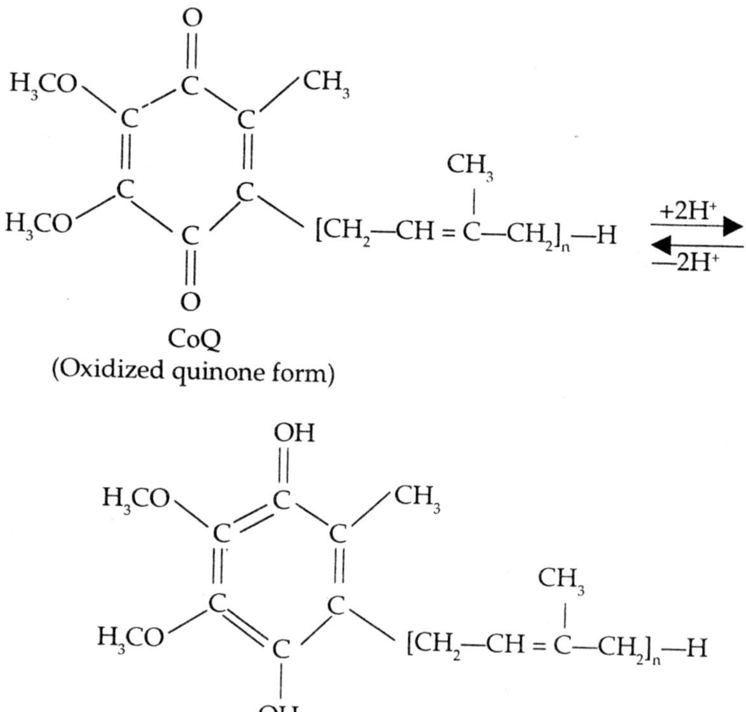

CoQ
(Oxidized quinone form)

$CoQH_2$
(Reduced hydroquinone form)

Fig.2. Oxidised and reduced forms of coenzyme Q.

Complex II: The second of the four membrane-bound complexes, succinate-CoQ oxidoreductase, also catalyzes the transfer of electrons to CoQ However, its source of electrons (in other words, the substance being oxidized) differs from the oxidizable substrate (NADH) acted on by NADHCoQ oxidoreductase, the first of the four respiratory complexes. In this case the substrate is succinate from the citric acid cycle, which is 'oxidized to fumarate by a flavin enzyme.

$$\text{Succinate} + \text{E-FAD} \rightarrow \text{fumarate} + \text{E-FADH}_2$$

The notation E-FAD indicates that the flavin portion is covalently bonded to the enzyme. The flavin group is reoxidized in the next stage of the reaction, and coenzyme Q is reduced.

$$\text{E-FADH}_2 + \text{CoQ} \rightarrow \text{E-FAD} + \text{CoQH}_2$$

The overall reaction is

$$\text{Succinate} + \text{CoQ} \rightarrow \text{fumarate} + \text{CoQH}_2$$

The other components of Complex II are a *b*-type cytochrome and two iron-sulphur proteins. The whole complex is an integral part of the inner mitochondrial membrane. The standard free energy change is -13.5 kj mol $^{-1}$ = -3.2 kcal mol $^{-1}$. The overall reaction is exergonic, but there is not enough energy from this reaction to drive ATP production.

In further steps of the electron transport chain, electrons are passed from coenzyme Q which is then reoxidized, to the first of a series of very similar proteins called cytochromes. Each of these proteins contains a heme group, and in each heme group the iron is successively reduced to Fe (II) and reoxidized to Fe (III). This situation differs from that of the iron in the heme group of hemoglobin, which remains in the reduced form as Fe (II) through the entire process of oxygen transport in the bloodstream. There are also some structural differences between the heme group in hemoglobin and the heme groups in the various types of cytochromes.

The successive oxidation-reduction reactions of the cytochromes

$$\text{Fe (III)} + \text{C} \rightarrow \text{Fe (II)} \text{ (reduction)}$$

and

$$Fe(II) \rightarrow Fe\ (III) + C\ (oxidation)$$

Differ from one another because the free energy of each reaction, $\Delta G^{\circ\prime}$, differs from the others because cf the influences of the various types of hemes and protein structure. Each of the proteins is slightly different in structure and thus each protein has slightly different properties, including the tendency to participate in oxidation-reduction reactions.

Complex III: The third complex, $CoQH_2$-Cytochrome c oxidoreductase (also called cytochrome reductase), catalyzes the oxidation of reduced coenzyme Q ($CoQH_2$). The electrons produced by this oxidation reaction are passed along to cytochrome c in a multistep process. The overall reaction is

$$CoQH_2 + 2\ \text{cyt}\ c[Fe\ (III)] \rightarrow CoQ + 2\ \text{cyt}\ c[Fe(II)] + 2H^+$$

The oxidation of coenzyme Q involves two electrons, whereas the reduction of Fe(III) to Fe(II) requires only one electron. Therefore, two molecules of cytochrome c are required for every molecule of coenzyme Q. The components of this complex include cytochrome *b* (actually two *b*-type cytochromes), cytochrome *c*, and several iron-sulphur proteins. The third complex is an integral part of the inner mitochondrial membrane.

Coenzyme Q is soluble in the lipid component of the mitochondrial membrane and is separated from the complex in the fractionation process that resolves the electron transport apparatus into its component parts, but the coenzyme is probably close to respiratory complexes in the intact membrane.

Cytochrome *c* itself is not part of the complex but is

loosely bound to the outer surface of the inner mitochondrial membrane, facing the intermembrane space. It is noteworthy that these two important electron carriers, coenzyme Q and cytochrome *c*, are not part of the respiratory complexes but can move freely in the membrane.

The respiratory complexes themselves move within the membrane, and electron transport occurs when one complex encounters the next complex in the respiratory chain as they move. The flow of electrons from reduced coenzyme Q to the other components of the complex does not take a simple direct path. It is becoming clear that a cyclic flow of electrons involves coenzyme Q twice. This behaviour depends on the fact that, as a quinone, coenzyme Q can exist in three forms.

The semiquinone form, which is intermediate between the oxidized and reduced forms, is of crucial importance here. Because of the crucial involvement of coenzyme Q, this portion of the pathway is called the Q cycle. In part of the Q cycle, one electron is passed from reduced coenzyme Q to the iron-sulphur clusters to cytochrome cl, leaving coenzyme Q in the semiquinone form.

$$COQH_2 \rightarrow Fe—S \rightarrow cyt\ c_1$$

The notation Fe-S indicates the iron-sulphur clusters. The series of reactions involving coenzyme Q and cytochrome c_1, but omitting the iron-sulphur proteins, can be written as follows:

$$CoQH_2 + \text{cytochrome } c_1 \text{ (oxidized)} \rightarrow \text{cytochrome } c_1 \text{ (reduced)} + CoQ\text{- (semiquinone anion)} + 2H^+$$

The semiquinone, along with the oxidized and reduced forms of coenzyme Q participates in a cyclic process in which the two b cytochromes are reduced and oxidized in turn. A second molecule of coenzyme Q is involved, transferring a second electron to cytochrome c_1, and from there to the mobile carrier cytochrome *c*.

We are going to omit a number of details of the process in the interest of simplicity. Each of the two molecules of coenzyme Q involved in the Q cycle loses one electron. The net result is the same as if one molecule of CoQ had lost two electrons. It is known that one molecule of $CoQH_2$ is regenerated, and one is oxidized to CoQ which is consistent with this picture.

Most importantly, the Q cycle provides a mechanism for electrons to be transferred one at a time from coenzyme Q to cytochrome c_1. Proton pumping, to which ATP production is coupled, occurs as a result of the reactions of this complex. The Q cycle is implicated in the process, and the whole topic is under active investigation. The standard free energy change (ΔG°¢) is -34.2 kj = - 8.2 kcal for each mole of NADH that enters the electron transport chain. The phosphorylation of ADP requires 30.5 kj mol^{-1} = 7.3 kcal mol^{-1}, and the reaction catalyzed by the third complex supplies enough energy to drive the production of ATP.

Complex IV: The fourth complex, cytochrome *c* oxidase, catalyzes the final steps of electron transport, the transfer of electrons from cytochrome *c* to oxygen. The overall reaction is

$$2 \text{ Cyt c[Fe(11)]} + 2H^+ + \tfrac{1}{2}O_2 \rightarrow \text{cyt c[Fe(III)]} + H_2O$$

Proton pumping also takes place as a result of this reaction. Like the other respiratory complexes, cytochrome oxidase is an integral part of the inner mitochondrial membrane and contains cytochromes *a* and a_3, as well as two CU^{2+} ions that are involved in the electron transport process. Taken as a whole, this complex contains about ten subunits. In the flow of electrons, the copper ions are intermediate electron acceptors that lie between the two *a*-type cytochromes in the sequence

$$\text{Cyt } c \rightarrow \text{cyt } a \rightarrow Cu^{2+} \rightarrow \text{Cyt } a_3 \rightarrow O_2$$

Cytochromes *a* and aa_3 taken together form the complex known as cytochrome oxidase. The reduced cytochrome oxidase is then oxidized by oxygen, which is itself reduced to water. The half reaction for the reduction of oxygen (oxygen acts as an oxidizing agent) is

$$\tfrac{1}{2}O_2 + 2H^+ + 2e \rightarrow H_2O$$

The standard free energy change ($\Delta G°'$) is -110 kj = -26.3 kcal for each mole of NADH that enters the electron transport chain. We have now seen the three places in the respiratory chain where electron transport is coupled to ATP production by proton pumping. These three places are the NADH dehydrogenase reaction, the oxidation of cytochrome *b*, and the reaction of cytochrome oxidase with oxygen.

In contrast to the electron carriers in the early stages of electron transport, such as NADH, FMN, and CoQ the cytochromes are macromolecules. These proteins are found in all types of organisms and are typically located in membranes. In eukaryotes the usual site is the inner mitochondrial membrane, but cytochromes can also occur in the endoplasmic reticulum. All cytochromes contain the heme group, a part of the structure of hemoglobin and myoglobin.

In the cytochromes the iron of the heme group does not bind to oxygen; instead, the iron is involved in the series of redox reactions, which we have already seen.

There are differences in the side chains of the heme group of the cytochromes involved in the various stages of electron transport.

These structural differences, combined with the variations in the polypeptide chain and in the way the polypeptide chain is attached to the heme, account for the differences in properties among the cytochromes in the electron transport chain. Non-heme iron proteins do not contain a heme group, as their name indicates. Many of the most important proteins in this category contain sulphur, as is the case with the iron-sulphur proteins that are components of the respiratory complexes. The iron is usually bound to cysteine or to S^{2-}. There are still many questions about the location and mode of action of iron-sulphur proteins in mitochondria.

Coupling of oxidation to phosphorelation

Some of the energy released by the oxidation reactions in the electron transport chain is used to drive the phosphorylation of ADP. The phosphorylation of each mole of ADP requires 30.5 kj = 7.3 kcal, and we have seen how each of the reactions catalyzed by three of the four respiratory complexes provides more than enough energy to drive this reaction.

It is a common theme in metabolism that energy to be used by cells is converted to the chemical energy of ATP as needed. The energy-releasing oxidation reactions give rise to proton pumping and thus to the pH gradient across the inner mitochondrial membrane.

In addition to the pH gradient, there is a voltage difference across the membrane generated by the concentration differences of ions inside and out. The energy of the electrochemical potential (voltage drop) across the membrane is converted to the chemical energy of ATP by the coupling process. Phosphorylation is not the same as electron transport, even though the two are intimately associated.

A coupling factor is needed to link oxidation and phosphorylation. A complex protein oligomer, separate from the electron transport complexes, serves this function; the complete protein spans the inner mitochondrial membrane and projects into the matrix as well. The portion of the protein that spans the membrane is called F_0. It consists of four different kinds of polypeptide chains, and research is in progress to characterize it further. The portion that projects into the matrix is called F_1; it consists of five different kinds of polypeptide chains in the ratio $\alpha_3\beta_3\gamma\delta\varepsilon$.

Electron micrographs of mitochondria show the projections into the matrix from the inner mitochondrial membrane. The F_1 sphere is the site of ATP synthesis. The whole protein complex is called ATP synthase. It is also known as mitochondrial ATPase, because the reverse reaction of ATP hydrolysis, as well as phosphorylation, can be catalyzed by the enzyme. The hydrolytic reaction was discovered before the reaction of the synthesis of ATP, hence the name. The 1997 Nobel Prize in Chemistry was shared by an American scientist, Paul Boyer of UCLA, and by a British scientist, John Walker of the Medical Research Council in Cambridge, England, for their elucidation of the structure and mechanism of this enzyme.

Compounds known as uncouplers inhibit the phosphorylation of ADP without affecting electron transport. A well-known example of an uncoupler is 2,4-dinitrophenol. Various antibiotics such as valinomycin and gramicidin A are also uncouplers. When mitochondrial oxidation processes are operating normally, electron transport from NADH or $FADH_2$ to oxygen results in the production of ATP. When an uncoupler is present, oxygen is still reduced to H_2O, but ATP is not produced.

If the uncoupler is removed, ATP synthesis linked to electron transport resumes. The coupling of electron transport to oxidative phosphorylation requires a multisubunit membrane-bound enzyme, ATP synthase. This enzyme has a hannel for protons to flow from the intermembrane space into the mitochondrial matrix. the proton flow is coupled to ATP production in a process that appears to involve a con:ormational change of the enzyme. A term called the P/O ratio is used to indicate the coupling of ATP production to electron transport.

The P/O ratio gives the number of moles of Pi consumed in the reaction ADP + Pi $\rightarrow$ ATP for each mole of oxygen atoms consumed in the reaction $O_2 + 2H^+ + 2e \rightarrow H_2O$. 2.5 moles of ATP are produced when one mole of NADH is oxidized to NAD^+ Recall that oxygen is the ultimate acceptor of the electrons from NADH and that 1 mole of O_2 molecules (one mole of oxygen atoms) is reduced for each mole of NADH oxidized. The experimentally determined P/O ratio is 2.5 when NADH is the substrate oxidized. The P/O ratio is 1.5 when $FADH_2$ is the substrate oxidized (also an experimentally determined value). Until recently, biochemists had used integral values of 3 and 2 for the P/O ratios for reoxidation of NADH and $FADH_2$, respectively. The nonintegral consensus values given here clearly underscore the complexity of electron transport, oxidative phosphorylation, and the manner in which they are coupled.

Copupling in oxidative phosphorylation

Several mechanisms have been proposed to account for the coupling of electron transport and ATP production. The mechanism that served as the point of departure in all discussions is chemoiosmotic coupling, which was later modified to include a consideration of conformational coupling.

Chemoiosmotic coupling

The chemoiosmotic coupling mechanism was based entirely on the difference in proton concentration between the intermembrane space and the matrix of an actively respiring mitochondrion. In other words, the proton (hydrogen ion, H^+) gradient across the inner mitochondrial membrane is the crux of the matter. The proton gradient exists because the various proteins that serve as electron carriers in the respiratory chain are not symmetrically oriented with respect to the two sides of the inner mitochondrial membrane, nor do they react in the same way with respect to the matrix and the intermembrane space.

In the process of electron transport the proteins of the respiratory complexes take up protons from the matrix to transfer them in redox reactions; these electron carriers subsequently release protons into the intermembrane space when they are reoxidized, creating the proton gradient. The reactions of the respiratory complexes all require protons, which come from the matrix side of the membrane.

The respiratory complexes transfer the protons to the intermembrane space. As a result, there is a higher concentration of protons in the intermembrane space than in the matrix, which is precisely what we mean by a proton gradient. It is known that the intermembrane space has a lower pH than the matrix, which is another way of saying that there is a higher concentration of protons in the intermembrane space than in the matrix. The proton gradient in turn can drive the production of ATP that occurs when the protons flow back into the matrix.

Chemiosmotic coupling was first suggested by the British scientist Peter Mitchell in 1961. A considerable body of experimental evidence has accumulated to support it.

1. A system with definite inside and outside compartments (closed vesicles) is essential for oxidative phosphorylation. The process does not occur in soluble preparations or in membrane fragments without compartmentalization.
2. Submitochondrial preparations that contain closed vesicles can be prepared; such vesicles can carry out oxidative phosphorylation, and the asymmetrical orientation of the respiratory complexes with respect to the membrane can be demonstrated.

A model system for oxidative phosphorylation can be constructed with proton pumping in the absence of electron transport. The model system consists of reconstituted membrane vesicles, mitochondrial ATP synthase, and a proton pump. The pump is bacteriorhodopsin, a protein found in the purple membrane of halobacteria. The proton pumping takes place when the protein is illuminated. The existence of the pH gradient has been demonstrated and confirmed experimentally. The way in which the proton gradient leads to the production of ATP depends on ion channels through the inner mitochondrial membrane; these channels are a feature of the structure of ATP synthase.

Protons flow back into the matrix through ion channels in the ATP synthase; the FO part of the protein is the proton channel. The flow of protons is accompanied by formation of ATV, which takes place in the F, unit. The unique feature of cherniosmotic coupling is the direct linkage of the proton gradient to the phosphorylation reaction.

The details of the way in which phosphorylation takes place as a result of the linkage to the proton gradient are not explicitly specified in this mechanism. A reasonable mode of action for uncouplers can be proposed in light of the existence of a proton gradient.

Dinitrophenol is an acid; its conjugate base, dinitrophenolate anion, is the actual uncoupler, since it can react with protons in the intermembrane space, reducing the difference in proton concentration between the two sides of the inner mitochondrial membrane. The antibiotic uncouplers such as gramicidin A and valinomycin are ionophores, creating a channel through which ions such as H^+, K^+ and Na^{2+} can pass through the membrane. The proton gradient is overcome, resulting in the uncoupling of oxidation and phosphorylation.

Conformational coupling

In conformational coupling the proton gradient is indirectly related to ATP production. The proton gradient leads to conformational changes in a number of proteins, particularly in the ATP synthase itself. It appears from recent evidence that the proton gradient is involved in the release of tightly bound ATP from the synthase as a result of the conformational change.

There are three sites for substrate on the synthase and three possible conformational states: open (O), with low affinity for substrate; loosebinding (L), which is not catalytically active; and tight-binding (T), which is catalytically active. At any given time, each of the sites is in one of three different conformational states.

These states interconvert as a result of the proton flux through the synthase. ATP already formed by the synthase is bound at a site in the T conformation, while ADP and Pi bind at a site in the L conformation. A proton flux converts the site in the T conformation to the

O conformation, releasing the ATP. The site at which ADP and Pi are bound assumes the T conformation, which can then give rise to ATP.

More recently, it has been shown that the F_1 portion of ATP synthase acts as a rotary motor. The y-E domain is the rotor, turning within a stationary barrel of a domain consisting of the 8 subunit in association with the $\alpha_3\beta_3$ hexamer. The $\gamma\varepsilon$ domain is the rotating "shaft" that mediates the energy exchange between the proton flow at F_0 and ATP synthesis at F_1. Experimental evidence long supported the idea that conformational changes play a role in the coupling of oxidation and phosphorylation. The discovery of the role of the $\gamma\varepsilon$ domain of F_1 firmly establishes the point.

Respiratory inhibitors

If a pipeline is blocked, there will be a backup. Liquid will accumulate upstream of the blockage point, but there will be less liquid downstream. In electron transport the flow of electrons is from one compound to another rather than along a pipe, but the analogy of a blocked pipeline can be useful for understanding the workings of the pathway.

When a flow of electrons is blocked in a series of redox reactions, there will be an accumulation of reduced compounds before the blockage point in the pathway. Recall that reduction is a gain of electrons, and oxidation represents a loss of electrons. The compounds that come after the blockage point will be lacking electrons and will tend to be found in the oxidized form.

By using respiratory inhibitors we can gather additional evidence to establish the order of components in the electron transport pathway. The use of respiratory inhibitors to determine the order of the electron transport chain depends on determining the relative amounts of oxidized and reduced forms of the various electron carriers in intact mitochondria. The logic of the experiment can be seen from the analogy of the blocked pipe.

In this case the reduced form of the carrier upstream will accumulate because it cannot pass electrons farther in the chain. Likewise, the oxidized form of the carrier downstream (cytochrome *c* in our example) will also accumulate because the supply of electrons that it could accept has been cut off.

By use of careful techniques, intact mitochondria can be isolated from cells and can carry out electron transport if an oxidizable substrate is available. If electron transport in mitochondria occurs in the presence and absence of a respiratory inhibitor, there will be different relative amounts of oxidized and reduced forms of the electron carriers. The type of experiment done to determine the relative amounts of oxidized and reduced forms of electron carriers depends on the spectroscopic properties of these substances.

The oxidized and reduced forms of cytochromes can be distinguished from one another by spectroscopic methods. Specialized spectroscopic techniques exist to detect the presence of electron carriers in intact mitochondria. The individual types of cytochromes can be identified by the wavelength at which the peak appears, and the relative amounts can be determined from the intensities of the peaks. There are three sites in the electron transport chain at which inhibitors have an effect, and we shall look at some classic examples.

At the first site, barbiturates block the transfer of electrons from the flavoprotein NADH reductase to coenzyme Q. Rotenone is another inhibitor that is active at this site. The second site at which block age can occur is that of electron transfer involving the *b* cytochromes, coenzyme Q and cytochrome c_1; the classic inhibitor associated with this blockage is the antibiotic antimycin A.

More recently developed inhibitors that are active in this part of the electron transport chain include myxothiazol and 5-n-undecyl-6-hydroxy-4,7-dioxobenzothiazol (UHDBT). These compounds played a role in establishing the existence of the Q cycle. The third site subject to blockage is the transfer of electrons from the cytochrome aa_3 complex to oxygen. Several potent inhibitors operate at this site, such as cyanide (CN^-), azide (N_3), and carbon monoxide- (CO).

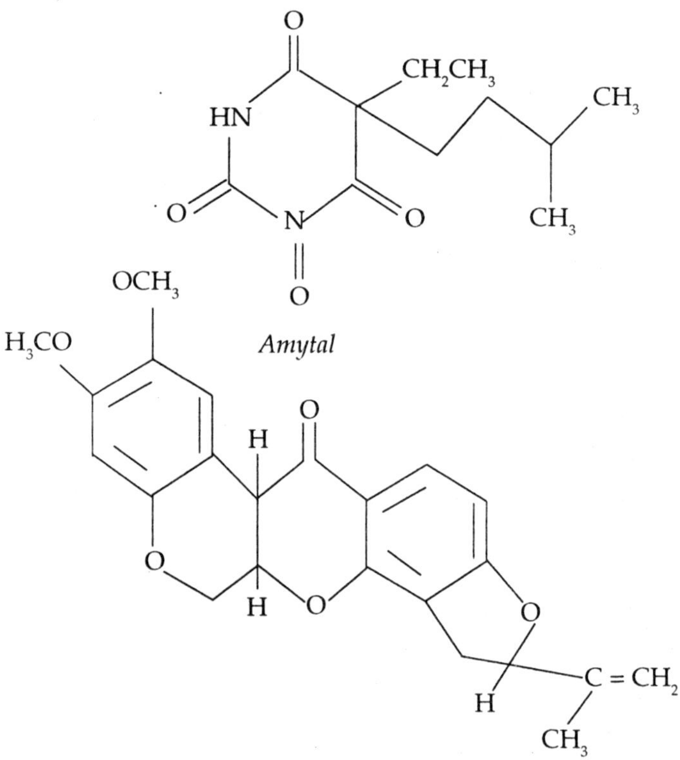

Fig. 3. Respiratory inhibitors

Each of the three sites of action of respiratory inhibitors corresponds to one of the respiratory complexes. Research is continuing with some of the more recently developed inhibitors; the goal of additional work is to elucidate more of the details of the electron transport process. Respiratory inhibitors block the electron transport chain at sites that correspond to each of the respiratory complexes. Experiments on these substances, many of which are highly toxic, were used to determine the path of electrons in respiration.

Shuttle mechanism

NADH is produced by glycolysis, which occurs in the cytosol, but NADH in the cytosol cannot cross the inner mitochondrial membrane to enter the electron transport chain. However, the electrons can be transferred to a carrier that can cross the membrane. The number of ATP molecules generated depends on the nature of the carrier, which varies according to the type of cell in which it occurs. One carrier system that has been extensively studied in insect flight muscle is the glycerol phosphate shuttle.

This mechanism makes use of the presence on the outer face of the inner mitochondrial membrane of an FAD-dependent enzyme that oxidizes glycerol phosphate. The glycerol phosphate is produced by the reduction of dihydroxyacetone phosphate; in the course of the reaction NADH is oxidized to NAD^+. In this reaction the oxidizing agent (which is itself reduced) is FAD, and the product is $FADH_2$. The $FADH_2$ then passes electrons through the electron transport chain, leading to the production of 1.5 moles of ATP for each mole of cytosolic NADH. This mechanism has also been observed in mammalian muscle and brain. A more complex and more efficient shuttle mechanism is the malateaspartate shuttle, which has been found in mammalian kidney, liver, and heart. This shuttle makes use of the fact that malate can cross the mitochondrial membrane, while oxaloacetate cannot do so.

The noteworthy point about this shuttle mechanism is that the transfer of electrons from NADH in the cytosol produces NADH in the mitochondrion. In the cytosol, oxaloacetate is reduced to malate by the cytosolic malate dehydrogenase, accompanied by the oxidation of cytosolic NADH to NAD^+. The malate then crosses the mitochondrial membrane.

In the mitochondrion, the conversion of malate back to oxaloacetate is catalyzed by the mitochondrial malate dehydrogenase (one of the enzymes of the citric acid cycle). Oxaloacetate is converted to aspartate, which can also cross the mitochondrial membrane. Aspartate is converted to oxaloacetate in the cytosol, completing the cycle of reactions. The NADH that is produced in the mitocondrion thus passes electrons to the electron transport chain.

Chapter 3

Carbohydrate Metabloism

When the word carbohydrate was coined, it originally referred to compounds of the general formula $Cn((H_2O)$. However, only the simple sugars, or monosaccharides, fit this formula exactly. The other types of carbohydrates, oligosaccharides and polysaccharides, are based on the monosaccharide units and have slightly different general formulas. Oligosaccharides are formed when a few (Greek oligos) monosaccharides are linked together; polysaccharides are formed when many (Greek polys) monosaccharides are bonded together. The reaction that adds monosaccharide units to a growing carbohydrate molecule involves the loss of one H_2O for each new link formed, accounting for the difference in the general formula.

Many commonly encountered carbohydrates are polysaccharides, including glycogen, which is found in animals, and starch and cellulose, which occur in plants. Carbohydrates play a number of important roles in biochemistry. First, they are major energy sources. Second, oligosaccharides play a key role in processes that take place on the surfaces of cells, particularly in cell-cell interactions and immune recognition. In addition, polysaccharides are essential structural components of several classes of organisms. Cellulose is a major component of grass and trees, and other polysaccharides are major components of bacterial cell walls.

Glucose is made by green plants and stored in the form of starch as the plants' energy reserves. Animals have an enzyme that recognizes the helical conformation of starch and can degrade it into its glucose units. Glucose, oxidized to carbon dioxide and water, is our primary energy source. Cellulose, a major component of plant cell walls and of cotton and wood, is a polymer of glucose monomers that all lie in the same plane. We don't possess the enzyme cellulase to break down cellulose, but termites do. A protozoan in their intestines contains cellulase, enabling them to digest wood. We learn this the hard way when wooden houses are damaged by termites. Modified carbohydrates are constituents of bacterial cell walls, where they are cross-linked by short polypepticle chains. By combining specific sugar monomers with amino acids, glycoproteins can be formed and act as cell surface markers to be recognized by other biomolecules.

Monosaccharides

A monosaccharide can be a polyhydroxy aldehyde (aldose) or a polyhydroxy ketone (ketose). The simplest monosaccharides contain three carbon atoms and are called trioses (tri meaning three). Glyceraldehyde is the aldose with three carbons (an aldotriose), and dihydroxyacetone is the ketose with three carbon atoms (a ketotriose). Aldoses with four, five, six, and seven carbon atoms are called aldotetroses, aldopentoses, aldohexoses, and aldoheptoses, respectively. The corresponding ketoses are ketotetroses, ketopentoses, ketohexoses, and ketoheptoses.

Six-carbon sugars are the most abundant in nature, but two five-carbon sugars, ribose and deoxyribose, occur in the structures of RNA and DNA, respectively. Four-carbon and seven-carbon sugars play roles in photosynthesis and other metabolic pathways.

Some molecules are not superimposable on their mirror images and that these mirror images are optical isomers (stereoisomers) of each other. A chiral (asymmetric) carbon atom is the usual source of optical isomerism, as was the case with amino acids. The simplest carbohydrate that contains a chiral carbon is glyceraldehyde, which can exist in two isomeric forms that are mirror images of each other.

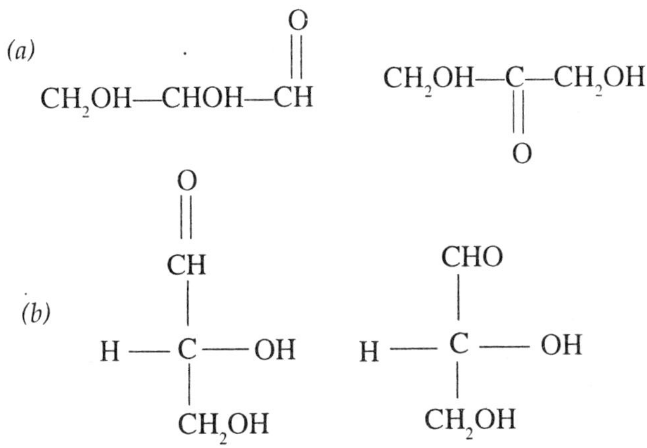

Fig. 1. Structure of simplest carbohydrates

The two forms differ in the position of the hydroxyl group bonded to the central carbon. (Dihydroxyacetone does not contain a chiral carbon atom and does not exist in nonsupefimposable mirror-image forms.) The two forms of glyceraldehyde are designated D-glyceraldehyde and L-glyceraldehyde. Mirror-image stereoisomers; are also called enantiomers, and D-glyceraldehyde and L-glyceraldehyde are enantiomers of each other.

Certain conventions are used for twodimensional drawings of the three-dimensional structures of stereoisomers. The dashed wedges represent bonds directed away from the viewer, below the plane of the paper, and the solid triangles represent bonds directed oppositely, toward the viewer and out of the plane of the paper. The configuration is the three-dimensional arrangement of groups around a chiral carbon atom, and stereoisomers differ from each other in configuration.

The two enantiomers of glyceraldehyde are the only possible stereoisomers of three-carbon sugars, but the possibilities for stereoisomerism increase as the number of carbon atoms increases. To show the structures of the resulting molecules, we need to say more about the convention for a twodimensional perspective of the molecular structure, which is called the Fischer projection method after the German chemist Emil Fischer, who established the structures of many sugars.

We shall use some common sixcarbon sugars for purposes of illustration. In the Fischer projection, bonds written "vertically" on the two-dimensional paper represent bonds directed behind the paper in three dimensions, whereas bonds written "horizontally" represent bonds directed in front of the paper in three dimensions.

The most highly oxidized carbon, in this case the one involved in the aldehyde group, is written at the "top" and is designated carbon 1, or C-1. The other carbon atoms are numbered in sequence from the "top." The designation of the configuration as L or D depends on-the arrangement at the chiral carbon with the highest number. In the cases of both glucose and fructose, this is C-5. In the Fischer projection of the D configuration, the hydroxyl group is on the right-of the highest-numbered chiral carbon, whereas in the L configuration the hydroxyl group is on the left of the highest-numbered chiral carbon.

Let us see what happens as another carbon is added to glyceraldehyde to give a four-carbon sugar. In other words, what are the possible stereoisomers for an aldotetrose? The aldotetroses have two chiral carbons, C-2 and C-3, and there are 22, or four, possible stereoisomers. Two of the isomers have the D configuration, and two have the L configuration. The two D isomers have the same configuration at C-3, but they differ in configuration (arrangement of the -OH group) at the other chiral carbon, C-2. These two isomers are called D-erythrose and D-threose. They are not superimposable on each other, but neither are they mirror images of each other.

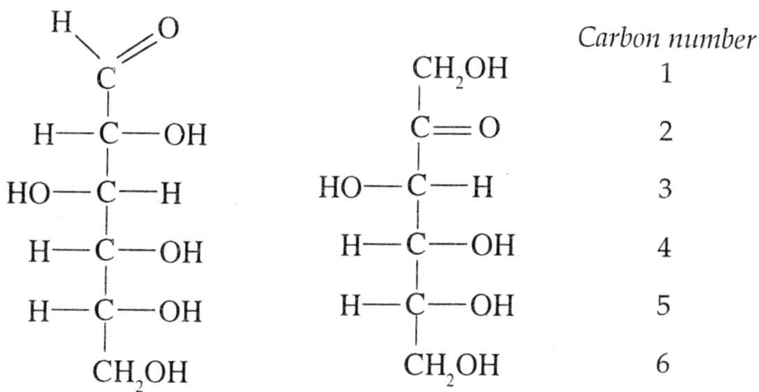

D-Glucose
(an aldose)

D-Fructose
(a ketose)

Fig. 2. Example of aldose and ketose

Such nonsuperimposable, non-mirror-image stereoisomers are called diastereomers. The two L isomers are L-erythrose and L-threose. L-Erythrose is the enantiomer (mirror image) of D-erythrose, and L-threose is the enantiomer of D-threose. L-Threose is a diastereomer of both D- and L-erythrose, and L-erythrose is a diastereomer of both D- and L-threose. Diastereomers that differ from each other in the configuration at only one chiral carbon are called epimers; D-erythrose and threose are epimers.

Aldopentoses have three chiral carbons, and there are 2 3, or eight, possible stereoisomers— four D forms and four L forms. Aldohexoses have four chiral carbons and 2 4, or sixteen, stereoisomers—eight D forms and eight L forms. Some of the possible stereoisomers are much more common in nature than others, and most biochemical discussion centres on the common naturally occurring sugars. For example, D sugars rather than L sugars predominate in nature. Most of the sugars we encounter in nature.

Biochemically important sugars usually contain five or six carbon atoms; their structure includes a carbonyl group and several hydroxyl groups. Optical isomerism is of paramount importance in the structure of simple sugars. Most of the important sugars found in nature have the D-Configuration, referred to the standard compound D-glyceraldehyde. Specially in foods, contain either five or six carbon atoms. We shall discuss D-glucose (an aldohexose) and D-ribose (an aldopentose) far more than many other sugars.

Anomers

Sugars, normally exist as cyclic molecules rather than as the open-chain forms we have shown so far. The cyclization takes place as a result of interaction between the functional groups on distant carbons such as C-1 and C-5 to form a cyclic hemiacetal (in aldohexoses). Another possibility is interaction between C-2 and C-5 to form a cyclic hemiketal (in ketohexoses). In either case, the carbonyl carbon becomes a new chiral center called the anomeric carbon.

The cyclic sugar can take either of two different forms, designated α and β called anomers of each other.

The Fischer projection of the α-anomer of a D sugar has the anomeric hydroxyl group to the right of the anomeric carbon (C-OH), and the P anomer of a D sugar has the anomeric hydroxyl group to the left of the anomeric carbon. The free carbonyl species can readily form either the α- or β- anomer, and the anomers can be converted from one form to another through the free carbonyl species.

In some biochemical reactions, any anomer of a given sugar can be used, but in other cases only one anomer occurs. For example, in living organisms only β-D-ribose and β–D-deoxyribose are found in RNA and DNA, respectively. Fischer projection formulas are useful for describing the stereochemistry of sugars, but their long bonds and right-angle bends do not give a realistic picture of the bonding situation in the cyclic forms, nor do they accurately represent the overall shape of the molecules.

a-configuration OH to right of anomeric carbon

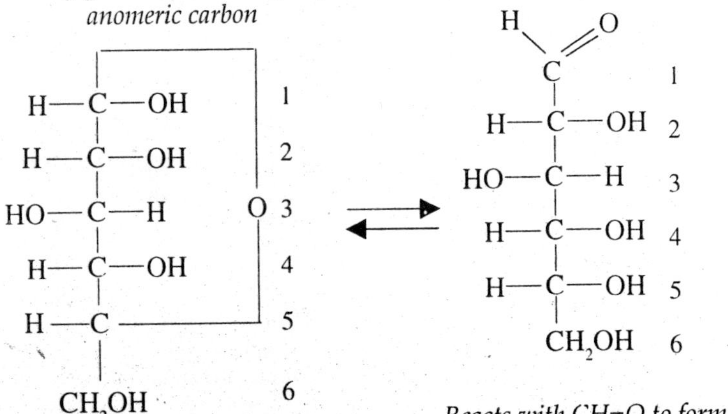

α-D-Glucose

Reacts with CH=O to form hemiacetal open chain form

Fig. 3. Fisher projection formulas of three forms of glucose.

Haworth projection formulas are more useful for those purposes. In Haworth projections, the cyclic structures of sugars are shown in perspective drawings as planar five- or sixmembered rings viewed nearly edge on. A five-membered ring is called a furanose because of its resemblance to furan; a six-membered ring is called a pyranose because of its resemblance to pyran.

These cyclic formulas approximate the shapes of the actual molecules better for furanoses than for pyranoses. The five-membered rings of furanoses are in reality very nearly planar, but the six-membered rings of pyranosres actually exist in solution in the chair conformation.

Even though the Haworth formulas are approximations, they are very useful because they are easy to draw. The Haworth projections represent the stereochemistry of sugars more realistically than do the Fischer projections, and the Haworth scheme is adequate for our purposes.

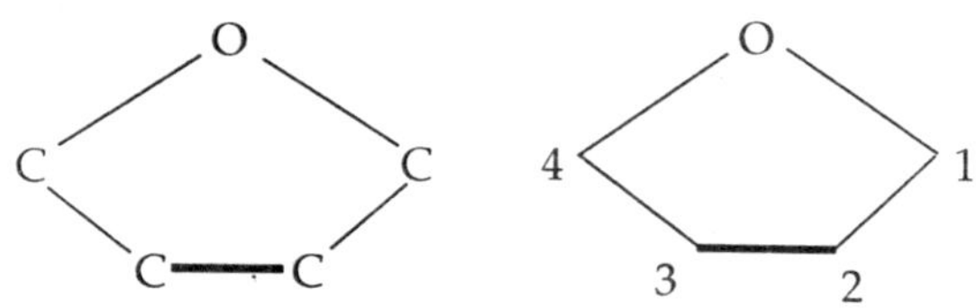

Haworth representations of furanose structures

Fig. 4. Haworth representations of sugar structures

We shall continue to use Haworth projections in our discussion of sugars. For a D sugar, any group that is written to the right of the carbon in a Fischer projection has a downward direction in a Haworth projection; any group that is written to the left in a Fischer projection has an upward direction in a Haworth projection. The terminal -CH_2OH group, which contains the carbon atom with the highest number in the numbering scheme, is shown in an upward direction.

Oxidation reduction reactions

Oxidation of sugars provides energy for organisms to carry out their life processes; the highest yield of energy from carbohydrates occurs when sugars are completely oxidized to CO_2 and H_2O in aerobic processes. The reverse of complete oxidation of sugars is the reduction of CO_2 and H_2O to form sugars, a process that takes place in photosynthesis. Several oxidation reactions of sugars are of some importance in laboratory practice, because they can be used for identification of sugars. Aldehyde groups can be oxidized to give the carboxyl group that is characteristic of acids, and this reaction is the basis of a test for the presence of aldoses.

When the aldehyde is oxidized, some oxidizing agent must be reduced. Aldoses are called reducing sugars because of this type of reaction; ketoses can also be reducing sugars because they isomerize to aldoses. In the cyclic form the compound produced by oxidation of an aldose is a lactone

Vitamin C (ascorbic acid) is an unsaturated lactone with a five-membered ring structure. Each carbon is bonded to a hydroxyl group except the carboxyl. carbon involved in the cyclic ester bond. Most animals can synthesize vitamin Q the exceptions are guinea pigs and primates, including humans. As a result, guinea pigs and primates must acquire vitamin C in their diet. Air oxidation of ascorbic acid, followed by hydrolysis of the ester bond, leads to loss of activity as a vitamin.

Consequently, lack of fresh food can cause vitamin C deficiencies, which in turn can lead to the disease scurvy . In this disease, defects in collagen structure cause fragile blood vessels. The presence of bydroxyproline is necessary for collagen stability because of hydrogen-bonded cross-links between collagen strands. Ascorbic acid, in turn, is essential for the activity of prolyl hydroxylase, which converts proline residues in collagen to hydroxyproline. Lack of ascorbic acid eventually leads to the fragile collagen responsible for the symptoms of scurvy. The British navy introduced citrus fruit into the diet of sailors in the 18th century to prevent scurvy during long sea voyages, and citrus fruit is still consumed by many for its vitamin C. Potatoes are another important source of vitamin C, not because potatoes contain a high concentration of ascorbic acid but because we eat so many potatoes.

Two types of reagent are used in the laboratory to detect the presence of reducing sugars. The first of these is Tollens' reagent, which uses the silver ammonia complex ion, $Ag\ (NH_3)^+{}_2$, as the oxidizing agent. A silver mirror is deposited on the wall of the test tube if a reducing sugar is present, as a result of the $Ag^+{}_2$ in the complex ion being reduced to free silver metal. A more recent method for detection of glucose, but not other reducing sugars, is based on the use of the enzyme glucose oxidase, which is specific for glucose.

In addition to oxidized sugars, there are some important reduced sugars. In deoxy sugars a hydrogen atom is substituted for one of the hydroxyl groups of the-sugar. One of these deoxy sugars is L-fucose (L-6-deoxygalactose), which is found in the carbohydrate portions of some glycoproteins. The name "glycoprotein" indicates that these substances are conjugated proteins that contain some carbohydrate group (glykos is Greek for "sweet") in addition to the polypeptide chain. An even more important example of a deoxy sugar is D-2-deoxyribose, the sugar found in DNA. When the carbonyl group of a sugar is reduced to a hydroxyl group, the resulting compound is one of the polyhydroxy alcohols known as alditols.

Esterification reactions

The hydroxyl groups of sugars behave exactly like all other alcohols in the sense that they can react with acids and derivatives of acids to form esters. The phosphate esters are particularly important ones because they are the usual intermediates in the breakdown of carbohydrates to provide energy. Phosphate esters are frequently formed by transfer of a phosphate group from ATP (adenosine triphosphate) to give the phosphorylated sugar and ADP (adenosine diphosphate).

Formation of glycosides

It is possible for a sugar hydroxyl group (ROH) bonded to the anomeric carbon to react with another hydroxyl (R'OH) to form a glycosidic linkage (W-O-R). A glycosidic linkage is not an ether (the W-O-R notation is misleading) because glycosides can behydrolyzed to the original alcohols. This type of reaction involves the anomeric carbon of the sugar in its cyclic form.

A hemiacetal carbon can react with an alcohol such as methyl alcohol to give a full acetal or glycoside. The newly formed bond is called a glycosidic bond. The glycosidic bonds discussed in this chapter are O-glycosides, with each sugar bonded to an oxygen atom of another molecule.

Glycosides derived from furanoses are called furanosides, and those derived from pyranoses are called pyranosides. Glycosidic bonds between monosaccharide units are the basis for the formation of oligosaccharides and polysaccharides. Glycosidic linkages can take various forms; the anomeric carbon of one sugar can be bonded to any one of the -OH groups on a second sugar to form an α or β glycosidic linkage.

Many different combinations are found in nature. The -OH groups are numbered so that they can be distinguished, and the numbering scheme follows that of the carbon atoms: The notation for the glycosidic linkage between the two sugars specifies which anomeric form of the sugar is involved in the bond and also specifies which carbon atoms of the two sugars are linked together. Two ways in which two α-D-glucose molecules can be linked together are a $(1\rightarrow 4)$ and α $(1\rightarrow 6)$.

In the first example, the α-anomeric carbon (C-1) of the first glucose molecule is joined in a glycosidic bond to the fourth carbon atom (C-4) of the second glucose molecule; the C-1 of the first glucose molecule is linked to the C-6 of the second glucose molecule in the second example.

When oligosaccharides and polysaccharides form as a result of glycosidic bonding, their chemical natures depend on which monosaccharides are linked together and also on the particular glycosidic bond formed. The difference between cellulose and starch depends on the glycosidic bond formed between glucose monomers.

Because of the variation in glycosidic linkages, both linear and branched-chain polymers can be formed. If the internal monosaccharide residues that are incorporated in a polysaccharide form two glycoside bonds, the polymer will be linear. Some internal residues can form three glycosidic bonds, leading to the formation of branched-chain structures. Another point about glycosides is worth mentioning.

The anomeric carbon is frequently involved in the glycosidic linkage, and also that the test for the presence of sugars,specifically for reducing sugars, requires a reaction of the group at the anomeric carbon. The internal anomeric carbons in oligosaccharides are not free to give the test for reducing sugars. Only if the end residue is a free hemiacetal rather than a glycoside will there be a positive test for a reducing sugar.

The level of detection can be important for such a test. A sample that contains only a few molecules of a large polysaccharide, each molecule with a single reducing end, might well produce a negative test because there are not enough reducing ends to detect.

In sucrose, and in starches and other sugar polymers, the olglycoside bonds attach sugars to sugars. Other major categories of glycosides are known in which the sugar binds to some other type of molecule. Probably the most common example is the structure of nucleotides, glycosides, in which the sugar binds to the nitrogenous, aromatic base, as found in ATP, many vitamins, and DNA and RNA. In glycolipids and glycoproteins, carbohydrates are attached to both lipids and proteins, respectively, by glycoside linkages. The red and blue colors of some flowers are sugar derivatives, often called anthocyanins.

These pigments involve various sugars bonded to the compound cyanidin and its derivatives. These compounds are water soluble because of the polar groups they possess. You may have done an acid-base titration of the pigment from red cabbage or from blue-berry-juice in a chemistry lab. In contrast, orange, yellow, and green plant pigments tend to be lipid in composition and insoluble in water. Many flavours involve sugar glycosides. Two familiar ones are cumamon and vanilla, in which the sugars bond to cinnamaldehyde (3-phenyl-2-propenal) and vanillin, respectively. Both of these compounds are aromatic aldehydes.

Many substances of medical importance have glycosidic linkages as part of their structure. Digitalis, prescribed for irregular heartbeat, is a mixture of several steroid complexes with sugars attached. Laetrile, a benzaldehyde derivative with a glycosidic linkage to glucuronic acid, was once thought to fight cancer, possibly since the cyanide moiety would poison the fast-growing cancer cells. This treatment is not approved in the United States, and it is likely that the cyanide causes more problems than it solves.

Other sugars derivatives

Amino sugars are an interesting class of compounds related to the monosaccharides. We shall not go into the chemistry of their formation, but it will be useful to have some acquaintance with them when we discuss polysaccharides. In sugars of this type, an amino group ($-NH_2$) or one of its derivatives is substituted for the hydroxyl group of the parent sugar. In N-acetyl amino sugars the amino group itself carries an acetyl group (CH_3-CO-) as a substituent.

Two particularly important examples are N-acetyl-β-glucosamine and its derivative N-acetyl-β-muramic acid, which has an added carboxylic acid side chain. These two compounds are components of bacterial cell walls. We did not specify whether N-acetylmuramic acid belongs to the L or the D series of configurations, and we did not specify the α- or β-anomer. This type of shorthand is the usual practice with α–β-glucose and its derivatives; the D configuration and the β-anomeric form are so common that we need not specify them all the time unless we want to make some specific point.

Oligosaccharides

Oligomers of sugars frequently occur as disaccharides,

formed by linking two monosaccharide units by glycosidic bonds. Three of the most important examples of oligosaccharides are disaccharides. They are sucrose, lactose, and maltose. Sucrose is the common table sugar extracted from sugarcane and sugar beets. The monosaccharide units that make up sucrose are α-D-glucose and b-D-fructose.

Glucose (an aldohexose) is a pyranose, and fructose (a ketohexose) is a furanose. The αC-1 carbon of the glucose is linked to the βC-2 carbon of the fructose in a glycosidic linkage that has the notation a, b(1→ 2). Sucrose is not a reducing sugar because both anomeric groups are involved in the glycosidic linkage. Free glucose is a reducing sugar, and free fructose can also give a positive test even though it is a ketone rather than an aldehyde in the open-chain form.

Fructose and ketoses in general can act as reducing sugars because they can isomerize to aldoses in a rather complex rearrangement reaction. When sucrose is consumed by animals, it is hydrolyzed to glucose and fructose, which are then degraded by metabolic processes to provide energy. Humans consume large quantities of sucrose, and excess consumption can contribute to health problems, which fact has led to a search for other sweetening agents. One that has been proposed is fructose itself. It is sweeter than sucrose, and therefore a smaller amount (by weight) of fructose than sucrose can produce the same sweetening effect with fewer "calories."

High-fructose corn syrup is frequently used in food processing. Also, one frequently sees the term "invert sugar" on food labels. That expression refers to the mixture of glucose and fructose produced by the hydrolysis of sucrose. The presence of fructose changes the texture of food, and the reaction to the change tends to depend on the preference of the consumer.

Artificial sweeteners have been produced in the laboratory and have frequently been suspected of having harmful side effects; the ensuing controversies bear eloquent testimony to the human craving for sweets. Saccharin, for example, has been found to cause cancer in laboratory animals, as have cyclamates, but the applicability of these results to human carcinogenesis has been questioned by some. Aspartame has been suspected of causing neurological problems, especially in individuals whose metabolisms cannot tolerate phenylalanine. It is safe to predict that the search for nonfattening sweeteners will continue and that it will be accompanied by controversy.

Polysaccharides are formed by the linking of monomeric sugars through glycosidic linkages These polymers, such as cellulose, starch, and glycogen, play important roles in the structure of organisms, such as plants and bacteria, and as vehicles for energy storage. Lactose is a disaccharide made up of β-D-galactose and D-glucose. Galactose is an epimer of glucose.

In other words, the difference between glucose and galactose is inversion of configuration at C-4. The glycosidic linkage is β(1→ 4), between the anomeric carbon C-1 of the P form of galactose and the C-4 carbon of glucose. Since the anomeric carbon of glucose is not involved in the glycosidic linkage, it can be in either the a or P form. The two anomeric forms of lactose can be specified, and the designation refers to the glucose residue; galactose must be present as the β-anomer, since the P form of galactose is required by the structure of lactose.

Lactose is a reducing sugar because the group at the anomeric carbon of the glucose portion is not involved in glycosidic: linkage, so it is free to react with oxidizing agents. Maltose is a disaccharide obtained from the hydrolysis of starch. It consists of two residues of D-glucose in an $\alpha(1 \rightarrow 4)$ linkage. Maltose differs from cellobiose, a disaccharide that is obtained from the hydrolysis of cellulose, only in the glycosidic linkage. In cellobiose the two residues of D-glucose are bonded together in a $\beta(1 \rightarrow 4)$ linkage.

Mammals can digest maltose, but not cellobiose. Yeast, specifically brewer's yeast, contains enzymes that hydrolyze the starch in sprouted barley (barley malt) first to maltose and then to glucose, which is fermented in the brewing of beer. Maltose is also used in other beverages, such as malted-milk.

Humans can be intolerant of milk and milk products for several reasons. Sugar intolerance involves the inability either to digest or to metabolize certain sugars. This situation differs from a food allergy, which involves the immune response. A negative reaction to sugars in the diet usually involves intolerance, whereas proteins, including those found in milk, tend to cause allergies. Most sugar intolerance is due to missing and defective enzymes, so this is another example of inborn errors of metabolism. Lactose is sometimes referred to as milk sugar because it occurs in milk.

In some adults a deficiency of the enzyme lactase in the intestinal villi causes a buildup of the disaccharide when milk products are ingested. This is because the enzyme is necessary to degrade lactose to galactose and glucose so that it can be absorbed into the bloodstream from the villi. Without the enzyme, built-up lactose in the intestine can be acted on by the lactase of intestinal bacteria (as opposed to the desirable lactase of the villi), producing hydrogen gas, carbon dioxide, and organic acids.

The products of the bacterial lactase reaction lead to digestive problems such as bloating and diarrhea, as does the presence of undegraded lactose. In addition, the by-products of the extra bacterial growth draws water into the intestine, thus aggravating the diarrhea. This disorder affects only about one-tenth of the Caucasian population of the United States, but it is more common among African-Americans, Asians, Native Americans, and Hispanics. Even if the enzyme lactase is present so that lactose can be broken down by the body, other problems can occur. A different, but related, problem can occur in the further metabolism of galactose, which must be isomerized to glucose if it is to enter the usual metabolic pathways.

If the enzyme that catalyzes this reaction is missing and galactose builds up, a condition known as galactosemia can result. This is a severe problem in infants, since the nonmetabolized galactose accumulates within cells and is converted to the hydroxy sugar galactitol, which cannot escape. Water is drawn into these cells and the swelling and edema cause damage. The critical tissue is the brain, which is not fully developed at birth. The swelling cells crush the brain tissue, and severe and irreversible retardation occurs. The clinical test for this disorder is cheap and required in all states. The dietary therapy for these two problems is quite different. Lactose-intolerant individuals must avoid lactose throughout their lives. Fortunately, tablets like Lactaido are available to add to regular milk, as are lactose and galactose-free formulas for feeding infants.

True fermented food products such as yogurt and many cheeses (especially aged ones) have had their lactose degraded during fermentation. However, many foods are not processed in this way, so lactose-intolerant individuals need to exercise caution in their food choices. There is no treatment like Lactaie for galactose, so affected individuals must avoid milk during childhood.

Fortunately, a galactose-free diet is easy to achieve simply by avoiding milk. After puberty, other pathways for galactose alleviate the problem in most afflicted individuals.

Polysachharides

Polysaccharides that occur in organisms are usually composed of a very few types of monosaccharide components. A polymer that consists of only one type of monosaccharide is a homopolysaccharide; a polymer that consists of more than one type of monosaccharide is a heteropolysaccharide.

Glucose is the most common monomer. When there is more than one type of monomer, frequently only two types of molecules occur in a repeating sequence. A complete characterization of a polysaccharide includes specification of which monomers are present and, if necessary, the sequence of monomers, It also requires that the type of glycosidic linkage be specified.

Cellulose and starch

Cellulose is the major structural component of plants, especially of wood and plant fibers. It is a linear hornopolysaccharide of β-D-glucose, and all residues are linked in β(I → 4) glycosidic bonds. Individual polysaccharide chains are hydrogen bonded together, giving plant fibers their mechanical strength. Animals lack enzymes, called cellulases that hydrolyze cellulose to glucose. Such enzymes attack the β–linkage, which is common to structural polymers; the a-linkage, which animals can digest, is characteristic of energy-storage polymers such as starch.

Termites and cattle can digest the cellulose in both grass and wood because bacteria in their digestive tracks produce the enzyme cellulase, which hydrolyzes the klycosidic linkage in cellulose. Celulases are found in bacteria, including the bacteria that inhabit tile digestive tracts of insects such as termites and grazing animals such as cattle and horses. The presence of these bacteria explains when cows and horses can live on grass and hay but humans cannot. The damage done by termites to the wooden parts of buildings arises from their ability—because of the presence of suitable bacteria in their digestive tracts—to use cellulose in wood as a nutrient.

Starches

The importance of carbohydrates as energy sources suggest that there is some use for polysaccharides in metabolism. We shall now discuss in more detail some polysaccharides, such as starches, that serve as vehicles for storage of glucose. Starches are polymers of α-D-glucose that occur in plant cells, usually as starch granules in the cytosol. There is an a-linkage in starch, in contrast to the P-linkage in cellulose. The types of starches can be distinguished from one another by their degrees of chain branching. Amylose is a linear polymer of glucose, with all the residues linked together by $\alpha(1 \rightarrow 4)$ bonds. Amylopectin is a branched-chain polymer, With the branches starting at $\alpha(1 \rightarrow 6)$ linkages along the chain of $\alpha(1 \rightarrow 6)$ linkages.

The most usual conformation of amylose is a helix with six residues per turn. Iodine molecules can fit inside the helix to form a starch-iodine complex, which has a characteristic dark-blue colour. The formation of this complex is a well-known test for the presence of starch. If there is a preferred conformation for arnylopectin, it is not yet known. Since starches are storage molecules, there must be a mechanism for releasing glucose from starch when the organism needs energy. Both plants and animals contain enzymes that hydrolyze starches. Two of these enzymes, known as α and β-amylase, attack $\alpha(1 \rightarrow 4)$ linkages.

P-Amylase is an exoglycosidase that cleaves from the nonreducing end of the polymer. Maltose, a dimer of gincose, is the product of reaction. The other enzyme, α-amylase, is an endoglycosidase, which can hydrolyze a glycosidic linkage anywhere along the chain to produce glucose and maltose. Arnylose can be completely degraded to glucose and maltose by the two amylases, but arnylopectin is not completely degraded because the branching linkages are not attacked. There are, however, debranching enzymes that occur in both plants and animals; they degrade the α(1 → 6) linkages. When these enzymes are combined With the amylases, they contribute to the complete degradation of both forms of starch.

Glycogen

Although starches occur only in plants, in animals there is a similar carbohydrate storage polymer. Glycogen is a branched-chain polymer of α-D-glucose, and in this respect it is similar to the amylopectin fraction of starch. Like amylopectin, glycogen consists of a chain of α(1 → 4) linkages with α(1 → 6) linkages at the branch points. The main difference between glycogen and amylopectin is that glycogen is more highly branched. Branch points occur about every 10 residues in glycogen and about every 4 residues in amylopectin. Glycogen is found in animal cells in granules similar to the starch granules in plant cells.

Glycogen granules are observed in well-fed liver and muscle cells, but they are not seen in some other cell types, such as brain and heart cells under normal conditions. When the organism needs energy, various degradative enzymes remove glucose units. Glycogen phosphorylase is one such enzyme; it cleaves one glucose at a time from the nonreducing end of a branch to produce glucose-1-phosphate, which then enters the metabolic pathways of carbohydrate breakdown. Debranching enzymes also play a role in the complete breakdown of glycogen.

Dietry fiber

Fiber in the diet is colloquially called roughage. It is principally made of complex carbohydrates, may have some protein components, and is moderately to fully insoluble. The health benefits of fiber are just beginning to be fully realized. We have known for a long time that roughage stimulates peristaltic action and thus helps move the digested food through the intestines, decreasing the transit time through the gut.

Potentially toxic substances, in food and in bile fluid, bind to fiber and are exported from the body, thus preventing them from doing damage to the lower intestine or being reabsorbed there. Statistical evidence indicates that a high-fiber diet also reduces colon and other cancers, precisely because fiber binds suspected carcinogens. It is also plausible that the benefit is because of a ,lack of other items in the high-fiber diet: People on highfiber diets also tend to take in less fat and fewer calories. Any difference in heart disease or cancer may be caused by these other differences. There has been much publicity about fiber in the diet reducing cholesterol. Fiber does bind cholesterol and it certainly causes some decrease in the amount in the blood. The reduction, expressed as a percentage, is higher in cases where the original level of cholesterol is higher. There is, however, no definite evidence that lowering cholesterol using fiber will result in less heart disease.

Fiber comes in two forms: soluble and insoluble. The most common insoluble fiber is cellulose, as in lettuce, carrots, bean sprouts, celery, brown rice, most other vegetables, and many fruit skins- also pumpernickel bread. Insoluble fiber binds various molecules but otherwise merely forms bulk in the lower intestine. Soluble fibers include araylopectin and other pectins and complex starches. There is a higher proportion of this type of fiber in uncooked and mildly processed foods. Good sources of soluble fiber include bran (especially oat bran), barley, fresh fruits (with skins), brussels sprouts, potatoes with skin, beans, and* zucchini. Soluble fiber binds water very well, increasing satiety in the stomach.

Chitin

A polysaccharide that is similar to cellulose in both structure and function is chitin, which is also a linear homopolysaccharide with all the residues linked in $\alpha(1 \rightarrow 4)$ glycosidic bonds. Chitin differs from cellulose in the nature of the monosaccharide unit; in cellulose the monomer is β-D-glucose, and in chitin the monomer is N-acetyl- β-D-glucosamine. The latter compound differs from glucose only in the substitution of the N-acetylamino group ($-NH-CO-CH_3$) for the hydroxyl group (-OH) on carbon C-2. Like cellulose, chitin plays a structural role and has a fair amount of mechanical strength because the individual strands are held together by hydrogen bonds. It is a major structural component of the exoskeletons of invertebrates such as insects and crustaceans and it also occurs in cell walls of algae, fungi, and yeasts.

Polysaccharides and structure of cell walls

In organisms that have cell walls, such as bacteria and plants, the walls consist largely of polysaccharides. The cell walls of bacteria and plants have biochemical differences, however, which serve to emphasize the differences between prokaryotes and eukaryotes.

Bacterial cell walls

Heteropolysaccharides are major components of bacterial

cell walls. A distinguishing feature of prokaryotic cell walls is that the polysaccharides are cross-linked by peptides. The repeating unit of the polysaccharide consists of two residues held together by $\beta(1 \rightarrow 6)$ glycosidic links, as was the case in cellulose and chitin. One of the two monomers is N-acetyl-D-glucosamine, which occurs in chitin, and the other monomer is N-acetylmuramic acid.

The structure of N-acetylmuramic acid differs from that of N-acetylglucosamine by the substitution of a lactic acid side chain (-O-CH(CH_3)-COOH) for the hydroxyl group (-OH) on carbon 3. N-acetylmuramic acid is found only in prokaryotic cell walls; it does not occur in eukaryotic cell walls. The cross-links in bacterial cell walls consist of small peptides. We shall use one of the best-known examples as an illustration. In the cell wall of the bacterium Staphylococcus aureus an oligomer of four amino acids (a tetramer) is bonded to N-acetylmuramic acid, forming a side chain. The tetrapeptides are themselves cross-linked by another small peptide, in this case consisting of five amino acids.

The carboxyl group of the lactic acid side chain of A-acetylmuramic acid forms a peptide bond with the N-terminal end of a tetrapepticle that has the sequence L-Ala-D-Gln-L-Lys-D-Ala. Recall that bacteria! cell walls are one of the few places where D-amino acids occur in nature. The occurrence of D-amino acids and N-acetylmuramic acid in bacterial cell walls but not in plant cell walls shows a biochemical as well as structural difference between prokaryotes and eukaryotes. The tetrapeptide forms two cross-links, both of them to a pentapeptide that consists of five glycine residues, $(Gly)_5$. The glycine pentamers form peptide bonds to the C-terminal end and to the side chain β-amino group of the lysine in the tetrapeptide. This extensive cross-linking produces a three-dimensional network of considerable mechanical strength, which is why bacterial cell walls are extremely difficult to disrupt. The material that results from the cross-linking of polysaccharides by peptides is a peptidoglycan, so named because it has both peptide and carbohydrate components.

Plant cell walls

The other important polysaccharide component found in plant cell walls is pectin, a polymer made up mostly of D-galacturonic acid, a derivative of galactose in which the hydroxyl group on carbon 6 has been oxidized to a carboxyl group. Pectin is extracted from plants because it has commercial importance in the food processing industry as a gelling agent in fruit preserves, jams and jellies. The major nonpolysaccharide component in plant cell walls, especially in woody plants, is lignin. Lignin is a polymer of coniferyl alcohol, and it is a very tough and durable material. Unlike bacterial cell walls, plant cell walls contain comparatively little peptide or protein.

Glycoproteins contain carbohydrate residues in addition to the polypeptide chain. Some of the most important examples of glycoproteins are involved in the immune response; for example, antibodies, which bind to and immobilize antigens, are glycoproteins. Carbohydrates also play an important role as antigenic determinants, the portions of an antigenic molecule that antibodies recognize and to which they bind. An example of the role of the oligosaccharide portion of glycoproteins as antigenic determinants is found in human blood groups. There are four human blood groups, A, B, AB, and O.

The distinctions between the groups depend on the oligosaccharide portions of the glycoproteins on the surfaces of the blood cells called erythrocytes. In all blood types the oligosaccharide portion of the molecule contains the sugar L-fucose, mentioned earlier in this chapter as an example of a deoxy sugar. N-acetylgalactosamine is found at the nonreducing end of the oligosaccharide in the type A blood-group antigen. In type B blood, α-D-galactose takes the place of N-acetylgalactosamine. In type O blood neither of these terminal residues is present, and in type AB blood both kinds of oligosaccharide are present.

Glycoproteins play an important role in eukaryotic cell membranes. The sugar portions are added to the protein as it passes through the Golgi on its way to the cell surface. Those glycoproteins with an extremely high carbohydrate content are classified as proteoglycans. Proteoglycans are constantly being synthesized and broken down. If there is a lack of the lysosomal enzymes that degrade them, proteoglycans. The structures of the blood group antigenic determinants accumulate, with tragic consequences.

Glycogen metabolism

When we digest a meal high in carbohydrates, we have a supply of glucose that exceeds our immediate needs. We store glucose as a polymer, glycogen, that is similar to the starches found in plants; glycogen differs from amylopectin only in the degree of chain branching. In fact, glycogen is sometimes called "animal starch" because of this similarity.

A look at the metabolism of glycogen will give us some insights into how glucose can be stored in this form and made available on demand. In the degradation of glycogen, several glucose residues can be released at one time, one from each end of a branch, rather than one at a time as would be the case in a linear polymer. This feature is useful to an organism in meeting short-term demands for energy by increasing the glucose supply as quickly as possible.

The structure of glycogen is optimized for its ability to store and deliver energy quickly and for the longest amount of time possible. The key to this optimization is the average chain length of the branches. If the average chain length were much greater or much shorter, glycogen would not be as efficient a vehicle for energy storage and release on demand.

Breakdown of glycogen

Glycogen is primarily found in liver and muscle. Release of glycogen in liver is triggered by low levels of glucose in blood. Liver glycogen breaks down to glucose-6-phosphate, which is hydrolyzed to give glucose. The release of glucose from the liver by this breakdown of glycogen replenishes the supply of glucose in the blood. In muscle, glucose-6-phosphate obtained from glycogen breakdown enters the glycolytic pathway rather than being exported to the bloodstream. Three reactions play roles in the conversion of glycogen to glucose-6-phosphate.

The enzyme that catalyzes the first of these reactions is glycogen phosphorylase; the second reaction is catalyzed by phosphoglucomutase.

Glycogen phosphorylase cleaves the $\alpha(1 \rightarrow 4)$ linkages in glycogen. Complete breakdown requires debranching enzymes that degrade the $\alpha(1 \rightarrow 4)$ linkages. Note that no ATP is hydrolyzed in the first reaction. In the glycolytic pathway we saw another example of phosphorylation of a substrate directly by phosphate without involvement of ATP.

The phosphorylation of glyceraidehyde-3-phosphate to 1,3-bisphosphoglycerate. This is an alternative mode of entry to the glycolytic pathway that "saves" one molecule of ATP for each molecule of glucose because it bypasses the first step in glycolysis. When glycogen rather than glucose is the starting material for glycolysis, there is a net gain of three ATP molecules for each glucose monomer, rather than two ATP as when glucose itself is the starting point.

The debranching of glycogen involves the transfer of a "limit branch" of three glucose residues to the end of another branch, where they are subsequently removed by glycogen phosphorylase. The same glycogen debranching enzyme then hydrolyzes the $\alpha(1 \rightarrow 6)$ glycosidic bond of the last glucose residue remaining at the branch point, When an organism needs energy quickly, glycogen breakdown is important.

Muscle tissue can mobilize glycogen more easily than fat and can do so anaerobically. It is only after glycogen supplies are depleted that fat burning becomes important. Some athletes, particularly long-distance runners, try to build up their glycogen reserves before a race by eating large amounts of carbohydrates. Some sources claim that this strategy works only for men, not for women.

Glycogen loading

Glycogen is the primary energy source for a resting muscle, which then starts working. The energy of ATP hydrolysis derived from glycogen breakdown is initially produced anaerobically, with the lactic acid product being processed back to glucose in the liver.

As an athlete gets well-conditioned, the muscle cells have more mitochondria, allowing for more aerobic metabolism and also for processing fats as well as carbohydrates for energy. The idea behind glycogen loading is that if there is more available glycogen, then a person can carry out anaerobic metabolism for a longer period of time when the muscle starts working.

Theoretical calculations estimate that it takes 8 to 12 minutes to use all the glycogen in the skeletal muscle. Allowing for loading of exfra glycogen, it might last 1/2 hour. There is evidence that in the well-conditioned athlete glycogen-may be used more slowly, since there is higher fat utilization.

Early loading methods involved glycogen depletion for three days by means of a high-protein diet and extreme exercise followed by loading from a high-carbohydrate diet and resting. This does yield a marked increase of glycogen, but some of it then gets stored in the heart and the practice actually stresses the heart muscle. There is clearly some danger here.

There are also dangers associated with a high-protein diet, since too much protein often leads to mineral imbalance, also stressing the heart and the kidneys. Simple loading without previous extreme depletion does increase glycogen, but not as much; however, this increase does not risk potential stress to the heart. Simple loading merely involves eating pasta, starch, and complex carbohydrate diets for a few days before the strenuous exertion.

Formation of glycogen from glucose

Formation of glycogen from glucose is not the exact reversal of the breakdown of glycogen to glucose. The synthesis of glycogen requires energy which is provided by the hydrolysis of a nucleoside triphosphate, UTP.

In the first stage of glycogen synthesis, glucose-l-phosphate (obtained from glucose-6-phosphate by an isomerization reaction) reacts with UTP to produce uridine diphosphate glucose and pyrophosphate (PPj). The enzyme that catalyzes this reaction is UDP-glucose Pyrophosphorylase. The exchange of one phosphoric anhydride bond for another has a free energy change close to zero. The release of energy comes about when the enzyme inorganic pyrophosphatase catalyzes the hydrolysis of pyrophosphate to two phosphates, a strongly exergonic reaction.

It is common in biochemistry to see the energy released by the hydrolysis of pyrophosphate combined with the free energy of hydrolysis of a nucleoside triphosphate. The coupling of these two exergonic reactions to a reaction that is not energetically favourable allows an otherwise endergonic reaction to take place. The supply of UTP is replenished by an exchange reaction with ATP, catalyzed by nucleoside phosphate kinase:

$$\text{UDP} + \text{ATP} \Leftrightarrow \text{UTP} + \text{ADP}$$

This exchange reaction makes the hydrolysis of any nucleoside triphosphate energetically equivalent to the hydrolysis of ATP. The addition of UDPG to a growing chain of glycogen is the next step in glycogen synthesis. Each step involves formation of a new $\alpha(1 \rightarrow 4)$ glycosidic bond in a reaction catalyzed by the enzyme glycogen synthase. This enzyme cannot simply form a bond between two isolated glucose molecules; it must add to an existing chain with $\alpha(1 \rightarrow 4)$ glycosidic linkages.

The initiation of glycogen synthesis requires a primer for this reason. The hydroxyl group of a specific tyrosine of the protein glycogenin serves this purpose. In the first stage of glycogen synthesis, a glucose residue is linked to this tyrosine hydroxyl and glucose residues are successively added to this first one. Synthesis of glycogen requires the formation of $\alpha(1 \rightarrow 6)$ as well as $\alpha(1 \rightarrow 4)$ glycosidic linkages.

A branching enzyme accomplishes this task. It does so by transferring a segment about seven residues in length from the end of a growing chain to a branch point where it catalyzes the formation of the required $\alpha(1 \rightarrow 4)$ glycosidic linkage. This enzyme has already catalyzed the breaking of an $\alpha(1 \rightarrow 4)$ glycosidic linkage in the process of transferring the oligosaccharide segment. Each transferred segment must come from a chain at least 11 residues long; each new branch point must be at least four residues away from the nearest existing branch point.

Control mechanisms

How does an organism ensure that glycogen synthesis and glycogen breakdown do not operate simultaneously? If this were to occur, the main result would be the hydrolysis of UTP, which would waste chemical energy stored in the phosphoric anhydride bonds. A major controlling factor lies in the behaviour of glycogen phosphorylase. This enzyme is not only subject to allosteric control; it also represents another control feature: covalent modification.

Glycogen phosphorylase enzyme is a dimer that exists in two forms, the inactive T form and the active R form. In the T form (and only in the T form) it can be modified by phosphorylation of a specific serine residue on each of the two subunits. The esterification of the serines to phosphoric acid is catalyzed by the enzyme phosphorylase kinase, the dephosphorylation is catalyzed by phosphoprotein phosphatase.

The phosphorylated form of glycogen phosphorylase is called phosphorylase *a*, and the dephosphorylated form is called phosphorylase *b*. Phosphorylase a does not respond to the allosteric effectors that control the behaviour of phosphorylase *b* and vice versa. Glucose is an allosteric inhibitor of phosphorylase *a*, whereas ATP and glucose-6-phosphate (G6P) are allosteric inhibitors of phosphorylase *b*; AMP is an activator of phosphorylase *b*.

These differences ensure that glycogen will be degraded when there is a need for energy and for ATP, as is the case with high [AMP], low [G6P], and low [ATP]. When the reverse is true (low [AMP], high [G6P], and high [ATP]), the need for energy and consequently for glycogen breakdown is less. "Shutting down" glycogen phosphorylase activity is the appropriate response. The combination of covalent modification and allosteric control of the process allows for a degree of fine tuning that would not be possible with either mechanism alone.

Hormonal control enters into the picture as well. When epinephrine is released from the adrenal gland in response to stress, it triggers a series of eventsthat suppress the activity of glycogen synthase and stimulate that of glycogen phosphorylase. The activity of glycogen synthase is subject to the same combination of allosteric control and covalent modification. The only difference is that the response to effectors is the opposite of that of glycogen phosphorylase. The same enzymes (phosphorylase kinase and phosphoprotein phosphatase) that modify glycogen phosphorylase perform the same functions with glycogen synthase.

The fact that two target enzymes, glycogen phosphorylase and glycogen synthase, are modified in the same way by the same enzymes links the opposing processes of synthesis and breakdown of glycogen even more intimately.

The modifying enzymes are themselves subject to covalent modification and allosteric control. This feature complicates the process considerably but adds the possibility of an amplified response to small changes in conditions. A small change in the concentration of an allosteric effector of a modifying enzyme can cause a large change in the concentration of an active, modified target enzyme; this amplification response is because the substrate for the modifying enzyme is itself an enzyme.

At this point the situation has become very complex indeed, but it is a good example of how opposing processes of breakdown and synthesis can be controlled to the advantage of an organism. When we see in the next section how glucose is synthesized from lactate, we shall have another example, one that we can contrast with glycolysis, to explore in more detail how carbohydrate metabolism is controlled. The conversion of pyruvate to glucose occurs by a process called gluconeogenesis.

Gluconeogenesis is not the exact reversal of glycolysis. We first met pyruvate as a product of glycolysis, but it can arise from other sources to be the starting point of the anabolism of glucose. Some of the reactions of glycolysis are essentially irreversible; these reactions are bypassed in gluconeogenesis.

An analogy is that of a hiker who goes directly down a steep slope but who climbs back up the hill by an alternative, easier route. We shall see that the biosynthesis and the degradation of many important biomolecules follow different pathways. There are three irreversible steps in glycolysis, and it is in these three reactions that the differences between glycolysis and gluconeogenesis are found.

The first is the production of pyruvate (and ATP) from phosphoenolpyruvate. The second is the production of fructose-1,6-bisphosphate from fructose-6-phosphate, and the third is the production of glucose-6-phosphate from glucose. The first of these reactions is exergonic, and the reverse reaction is endergonic. Reversing the second and third reactions requires the production of ATP from ADP, which is also an endergonic reaction. The net result of gluconeogenesis is the reversal of these three glycolytic reactions, but the pathway is different, with different -reactions and different enzymes.

The conversion of pyruvate to phosphoenol-pyruvate in gluconeogenesis takes place in two steps. The first step is the reaction of pyruvate and carbon dioxide to give oxaloacetate. This step requires energy, which is available from the hydrolysis of ATP.

$$\text{Pyruvate} + \text{ATP} + CO_2 + H_2O \rightarrow \text{oxaloacetate} + \text{ADP} + \text{Pi} + 2\text{H}$$

The enzyme that catalyzes this reaction is pyruvate carboxylase, an allosteric enzyme. Acetyl-CoA is an allosteric effector that activates pyruvate carboxylase. This reaction is a control point in the pathway because the enzyme that catalyzes it is subject to allosteric regulation. If high levels of acetyl-CoA are present (in other words, if there is more acetyl-CoA than is needed to supply citric acid cycle), pyruvate (a precursor of acetyl-CoA) can be diverted to gluconeogenesis. (Oxaloacetate from the citric acid cycle can frequently be a starting point for gluconcogenesis as well.) Magnesium ion (Mg^{2+}) and biotin are also required for effective catalysis.

Biotin is a carrier of carbon dioxide; it has a specific site for covalent attachment of CO_2. The carboxyl group of the biotin forms an amide bond with the P-amino group of a specific lysine side chain of pyruvate carboxylase. The CO_2 is attached to the biotin, which in turn is covalently bonded to the enzyme, and then the CO_2 is shifted to pyruvate to form oxaloacetate

The conversion of oxaloacetate to phospho-enolpyruvate is catalyzed by the enzyme phosphoenolpyruvate carboxykinase. This reaction also involves hydrolysis of a nucleoside triphosphate, GTP in this case rather than ATP.

Oxaloacetate + GTP → phosphoenolpyruvate + CO_2 + GDP

The successive carboxylation and decarboxylation reactions are both close to equilibrium (they have low values of their standard free energies), and as a result, the conversion of pyruvate to phosphoenolpyruvate is also close to equilibrium (DG°′ = 2.1 kj mol 0.5 kcal mol-1). A small increase in the level of oxaloacetate can drive the equilibrium to the right, and a small increase in the level of phosphoenolpyruvate can drive it to the left.

A concept well known in general chemistry, the law of mass action, relates the concentrations of reactants and products in a system at equilibrium to the equilibrium constant. The concept of equilibrium also implies that a reaction will proceed to the right on addition of reactants and to the left on addition of products.

Pyruvate + ATP + GTP → phosphoenolpyruvate + ADP + GDP + Pi + $2H^+$

Role of sugar phosphates in gluconeogenesis

The other two reactions in which gluconeogenesis differs from glycolysis are ones in which a phosphate-ester bond to a sugar-hydroxyl group is hydrolyzed. Both reactions are catalyzed by phosphatases, and both reactions are exergonic. The first reaction is the hydrolysis of fructose-1,6-bisphosphate to produce fructose-6-phosphate and phosphate.

Fructose-1,6-bisphosphate → fructose-6-phosphate + Pi

This reaction is catalyzed by the enzyme fructose-1,6-phsphosphatase, an allosteric enzyme strongly inhibited by adenosine monophosphate (AMP) but stimulated by ATP. Because of allosteric regulation, this reaction is also a control point in the pathway. When the cell has an ample supply of ATP, the formation rather than the breakdown of glucose is favoured. The second reaction is the hydrolysis of glucose-6-phosphate to glucose and phosphate ion (DG°′ = -13.8 kj mol^{-1} = -3.3 kcal mol^{-1}) The enzyme that catalyzes this reaction is glucose-6-phosphatase.

$$\text{Glucose-6-phosphate} \rightarrow \text{glucose} + \text{Pi}$$

In glycolysis the phosphorylation reactions must be coupled to the hydrolysis of ATP to make them exergonic and thus energetically allowed. In gluconeogenesis the organism can make direct use of the fact that the hydrolysis reactions of the sugar phosphates are exergonic. The corresponding reactions are not the reverse of each other in the two pathways. They differ from each other in whether they require ATP and in the enzymes involved.

Control mechanism in carbon metabolism

Glucose is also the product of gluconeogenesis, which has the new effect of reversing glycolysis; glucose is also obtained from the breakdown of glycogen. The opposing pathways, glycolysis and gluconeogenesis on the one hand and the breakdown and synthesis of glycogen on the other hand are not the exact reversal of each other even though the net results are simply those of reversal. It is time to see how all these related pathways are controlled.

An important element in the control process involves fructose-2,6-bisphosphate (F2,6P). This compound is an important allosteric activator of phosphofructokinase, the key enzyme of glycolysis; it is also an inhibitor of fructose bisphosphate phosphatase, which plays a role in gluconeogenesis. A high concentration of F2,6P stimulates glycolysis, whereas a low concentration stimulates gluconeo genesis.

The concentration of F2,6P in a cell depends on the balance between its synthesis (catalyzed by phosphofructokinase-2 [PFK-2)) and its breakdown. The enzymes that control the formation and breakdown of F2,6P are themselves con trolled by a phosphorylation/ dephosphorylation mechanism similar to that we have already seen in the case of glycogen phosphorylase and glycogen synthase. Both enzyme activities are located on the same protein (a dimer of about 100 kD molecular weight).

Phosphorylation of the dimeric protein leads to an increase in activity of FBPase-2 and a decrease in the concentration of F2,6P, ultimately stimulating gluconeogenesis. Dephosphorylation of the dimeric protein leads to an increase in PFK-2 activity and an increase in the concentration of F2,6P, ultimately stimulating glycolysis.

A number of control mechanisms operate in carbohydrate metabolism. In the mechanism of substrate cycling, the synthesis and the breakdown of a given compound are catalyzed by two different enzymes. Energy is required, but independent control can be exercised over the two opposing processes.

Substrate cycling

The term substrate cycling refers to the fact that opposing reactions can be catalyzed by different enzymes. Consequently the opposing reactions can be independently regulated and have different rates. It would not be possible to have different rates with the same enzyme, because a catalyst speeds up a reaction and the reverse of the reaction at the same rate. We shall use the conversion of fructose-6-phosphate to fructose-1,6-bisphosphate and then back to fructose-6-phosphate as an example of a substrate cycle.

In glycolysis, the reaction catalyzed by phosphofructokinase is highly exergonic under physiological conditions.

$$\text{Fructose-6-phosphate} + \text{ATP} \rightarrow \text{fructose 1-6-bisphosphate} + \text{ADP}$$

The opposing reaction, which is part of gluconeogenesis, is also exergonic and is catalyzed by another enzyme, namely fructose-1,6-bisphosphatase.

$$\text{Fructose-1,6-bisphosphate} + H_2O \rightarrow \text{fructose-6-phosphate} + \text{Pi}$$

The opposing reactions are not the exact reverse of one another. If we add the two opposing reactions together, we obtain the net reaction

$$\text{ATP} + H_2O \rightarrow \text{ADP} + \text{Pi}$$

Hydrolysis of ATP is the energetic price that is paid for independent control of the opposing reactions. Using combinations of these control mechanisms, an organism can set up a division of labour among tissues and organs to maintain control of glucose metabolism. A particularly clear example is found in the Cori cycle, and we shall discuss it here. The Cori cycle is named for Gerty and Carl Cori, who first described it. There is cycling of glucose caused by glycolysis in muscle and gluconeogenesis in liver.

Glycolysis in skeletal muscle produces lactate under conditions of oxygen debt, such as during a sprint. Skeletal muscle has comparatively few mitochondria, so metabolism is largely anaerobic in this tissue. The buildup of lactate is responsible for the muscular aches that follow strenuous exercise. Gluconeogenesis recycles the lactate that is produced (lactate is first oxidized to pyruvate).

The process occurs to a great extent in the liver after the lactate is transported there by the blood. Glucose produced in the liver is transported back to skeletal muscle by the blood, where it becomes an energy store for the next burst of exercise. Note that we have here a division of labour between two different types of organs, muscle and liver. In the same cell (of whatever type) these two metabolic pathways, glycolysis and gluconeogenesis, are not highly active simultaneously.

When the cell needs ATP, glycolysis is more active; when there is little need for ATP, gluconeogenesis is more active. Because of the hydrolysis of ATP and GTP in the reactions of gluconeogenesis that differ from those of glycolysis, the overall pathway from two molecules of pyruvate back to one molecule of glucose is exergonic.

The conversion of pyruvate to lactate is exergonic, which means that the reverse reaction is endergonic. The energy released by the exergonic conversion of pyruvate to glucose by gluconeogenesis facilitates the endergonic conversion of lactate to pyruvate. Note that the Cori cycle requires the net hydrolysis of two ATP and two GTP. While ATP is produced by the glycolytic part of the cycle, the portion involving gluconeogenesis requires yet more ATP in addition to GTP.

Glycolysis:

$$\text{Glucose} + 2\ NAD^+ + 2\ ADP + 2\ Pi \rightarrow 2\ \text{pyruvate} + 2\ NADH + 4\ H^+ + 2\ ATP + 2H_2O$$

Gluconeogenesis

$$2\ \text{Pyruvate} + 2NADH + 4H^+ + 4\ ATP + 2GTP + 6H_2O \rightarrow$$

$$\text{glucose} + 2NAD^+ + 4\ ADP + 2GDP + 6Pi$$

Overall:

$$2ATP + 2GTP + 4H_2O \rightarrow 2ADP + 2GDP + 4\ Pi$$

The hydrolysis of both ATP and GTP is the price of increased simultaneous control of the two opposing pathways.

Pentose phosphate pathway

The pentose phosphate pathway is an alternative to glycolysis and differs from it in several important ways. In glycolysis, one of our most important concerns was the production of ATP. In the pentose phosphate pathway, the production of ATP is not the crux of the matter. As the name of the pathway indicates, five-carbon sugars, including ribose, are produced from glucose. Ribose plays an important role in the structure of nucleic acids.

Another important facet of the pentose phosphate pathway is the production of NADPH, a compound that differs from NADH by having one extra phosphate group esterified to the ribose ring of the adenine nucleotide portion of the molecule.A more important difference is the way these two coenzymes function. NADH is produced in the oxidative reactions that give rise to ATP. NADPH is a reducing agent in biosynthesis, which by its very nature is a reductive process.

The pentose phosphate pathway begins with a series of oxidation reactions that produce NADPH and five-carbon sugars. The remainder of the pathway involves nonoxidative reshuffling of the carbon skeletons of the sugars involved. The products of these nonoxidative reactions include substances such as fructose-6-phosphate and glyceraldehyde-3-phosphate, which play a role in glycolysis.

Oxidative reactions

In the first reaction of the pathway, glucose-6-phosphate is oxidized to 6-phosphoglucono-δ-lactone. The enzyme that catalyzes this reaction is glucose-6-phosphate dehydrogenase. Note that NADPH is produced by the reaction. The same sort of reaction takes place in the oxidation of sugars by metal ions. The hemiacetal form of the sugar is a cyclic structure resulting from the addition of one of the hydroxyl groups to the aldehyde group. It is oxidized to a lactone, which is a cyclic ester formed between a carboxylic acid group and an alcohol group elsewhere in the molecule. The cyclic ester bond of 6-phosphoglucono-δ-lactone is hydrolyzed in the next reaction. The open-chain compound 6-phospho-gluconate (the gluconic acid is present in its ionized form) is the product of this reaction, which is catalyzed by the enzyme lactonase.

The next reaction is an oxidative decarboxylation, and NADPH is produced once again. The 6-phosphogluconate molecule loses its carboxyl group, which is released as carbon dioxide, and the five-carbon keto-sugar (ketose) ribulose-5-phosphate is the other product. The enzyme that catalyzes this reaction is 6phosphogluconate dehydrogenase. Note that in the process the C-3 hydroxyl group of the 6-phosphogluconate is oxidized to form a α-keto acid, which is unstable and readily decarboxylates to form ribulose-5-phosphate.

Nonoxidative reactions

In the remaining steps of the pentose phosphate pathway, several reactions involve transfer of two- and three-carbon units. In order to keep track of the carbon backbone of the sugars and their aldehyde and ketone functional groups, we shall write the formulas in the open-chain form. There are two different reactions in which ribulose-5-phosphate isomerizes. In one of these reactions, catalyzed by phosphopentose 3-epimerase, there is an inversion of configuration around carbon atom 3, producing xylulose-5phosphate, which is also a ketose.

The other isomerization reaction, catalyzed by phosphopentose isomerase, produces a sugar with an aldehyde group (an aldose) rather than a ketone. In this second reaction, ribulose-5-phosphate isomerizes to ribose-5-phosphate. Ribose-5-phosphate is a necessary building block for the synthesis of nucleic acids and coenzymes such as NADH.

The group-transfer reactions that link the pentose phosphate pathway with glycolysis require the two 5-carbon sugars produced by the isomerization of ribulose-5-phosphate. Two molecules of xylulose-5-phosphate and one molecule of ribose-5-phosphate rearrange to give two molecules of fructose-6-phosphate and one molecule of glyceraidehyde-3-phosphate.

In other words, three molecules of pentose (with five carbon atoms each) give two molecules of hexose (with six carbon atoms each) and one molecule of a triose (with three carbon atoms). The total number of carbon atoms (15) does not change, but there is considerable rearrangement as a result of group transfer. Two enzymes, transketolase and transaldolase, are responsible for the reshuffling of the carbon atoms of sugars such as ribose-5-phosphate and xylulose-5-phosphate in the remainder of the pathway, which consists of three reactions.

Transketolase transfers a two-carbon unit, and transaldolase transfers a three-carbon unit. Transketolase catalyzes the first and third reactions in the rearrangement process, and transaldolase catalyzes the second reaction. In the first of these reactions a two-carbon unit from xylulose-5-phosphate (five carbons) is transferred to ribose-5-phosphate (five carbons) to give sedoheptulose-7-phosphate (seven carbons) and glyceraldehyde-3-phosphate (three carbons). This reaction is catalyzed by transketolase. In the reaction catalyzed by transaldolase, a three-carbon unit is transferred from the seven-carbon sedoheptulose-7-phosphate to the three-carbon glyceraldehyde-3-phosphate. The products of the reaction are fructose-6-phosphate (six carbons) and erythrose-4-phosphate (four carbons).

In the final reaction of this type in the pathway, xylulose-5-phosphate reacts with erythrose-4-phosphate. This reaction is catalyzed by transketolase. The products of the reaction are fructose-6-phosphate and glyceraldehyde-3phosphate. In the pentose phosphate pathway, glucose-6-phosphate can be converted to fructose-6-phosphate and glyceraldehyde-3-phosphate by a means other than the glycolytic pathway.

For this reason the pentose phosphate pathway is also called the hexose monophosphate shunt, and this name is used in some texts. A major feature of the pentose phosphate pathway is the production of ribose-5-phosphate and NADPH. The control mechanisms of the pentose phosphate pathway can respond to the varying needs of organisms for either or both of these compounds.

Control of the pentose phosphate pathway

The reactions catalyzed by transketolase and transaldolase are reversible, which allows the pentose phosphate pathway to respond to the needs of an organism. The starting material, glucose-6-phosphate, will undergo different reaction's depending on whether there is a greater need for ribose-5-phosphate or for

NADPH. The operation of the oxidative portion of the pathway depends strongly on the organism's requirement for NADPH. The need for ribose-5-phosphate can be met in other ways, since ribose-5-phosphate can be obtained from glycolytic intermediates without the oxidative reactions of the pentose phosphate pathway.

If the organism needs more NADPH than ribose-5-phosphate, the reaction series goes through the complete pathway just discussed. The oxidative reactions at the beginning of the pathway are needed to produce NADPH. The net reaction for the oxidative portion of the pathway is

$$6 \text{ glucose-6-phosphate} + 12 \text{ NADP}^+ + 6H_2O \rightarrow$$

$$6 \text{ ribose-5-phosphate} + 6CO_2 + 12 \text{ NADPH} + 12 H^+$$

A clinical manifestation of an enzyme malfunction in the pentose phosphate pathway is discussed in the box.

Hemolytic anemia

The pentose phosphate pathway is the only source of NADPH in red blood cells, which as a result are highly dependent on the proper functioning of the enzymes involved. A glucose-6-phosphate dehydrogenase deficiency leads to an NADPH deficiency, which can in turn lead to hemolytic anaemia because of wholesale destruction of red blood cells. The relationship between NADPH deficiency and anemia is an indirect one. NADPH is required to reduce the peptide glutathione from the disulfide to the free thiol form.

Mammalian red blood cells lack mitochondria in which many redox reactions take place. Consequently, these cells are limited in the ways in which they can deal with redox balance. A substance like glutathione, which can take part in redox reactions, assumes greater importance than would be the case in cells with large numbers of mitochondria.

The presence of the reduced form of glutathione is necessary for the maintenance of the sulfhydryl groups of hemoglobin and other proteins in their reduced forms, as well as for keeping the Fe(II) of hemoglobin in its reduced form. Glutathione also maintains the integrity of red cells by reacting with peroxides that would otherwise degrade aminoacid side chains in the cell membrane. About 11% of African-Americans are affected by glucose-6-phosphate- dehydrogenase deficiency. This condition, like the sickle-cell trait, leads to increased resistance to malaria, accounting for some of its persistence in the gene pool in spite of its otherwise deleterious consequences.

Chapter 4

Glycolysis

In glycolysis, glucose is converted to pyruvate in a multistep pathway. When pyruvate is formed, it can be converted to carbon dioxide and water in aerobic reactions. It can also be converted to lactate under anaerobic conditions or, in some organisms, to ethyl alcohol. In glycolysis, one molecule of glucose (a six-carbon compound) is converted to fructose-1,6-bisphosphate (also a six-carbon compound), which eventually gives rise to two molecules of pyruvate (a three-carbon compound).

The glycolytic pathway (also called the Embden-Meyerhoff pathway) involves many steps, including the reactions in which metabolites of glucose are oxidized; there are other steps as well. Each reaction in the pathway is catalyzed by an enzyme specific for that reaction. In each of two reactions in the pathway, one molecule of ATP is hydrolyzed for each molecule of glucose metabolized; the energy released in the hydrolysis of these two ATP molecules makes coupled endergonic reactions possible. In each of two other reactions two molecules of ATP are produced by phosphorylation of ADP for each molecule of glucose, giving a total of four ATP molecules produced. A comparison of the number of ATP molecules used by hydrolysis (two) and the number produced (four) shows that there is a net gain of two ATP molecules for each molecule of glucose processed in glycolysis.

Glycolysis plays a key role in the way organisms extract energy from nutrients.When pyruvate is formed, it can have one of several fates. In aerobic metabolism (in the presence of oxygen), pyruvate loses carbon dioxide; the remaining two carbon atoms become linked to co-enzyme A as an acetyl group to form acetyl-CoA, which then enters the citric acid cycle.

There are two fates for pyruvate in anaerobic metabolism (in the absence of oxygen). In organisms capable of alcoholic fermentation, pyruvate loses carbon dioxide, this time producing acetaldehyde, which in turn is reduced to produce ethanol. The more common fate of pyruvate in anaerobic metabolism is reduction to lactate, called anaerobic glycolysis, to distinguish it from conversion of glucose to pyruvate, which is simply called, glycolysis.

Anaerobic metabolism is the only energy source in mammalian red blood cells, as well as in several species of bacteria, such as Lactobacillus in sour milk and Clostfidium botulinum in tainted canned foods. In all these reactions the conversion of glucose to product is an oxidation reaction, requiring an accompanying reduction reaction in which NAD is converted to NADH, a point to which we shall return when we discuss the pathway in detail.

The breakdown of glucose to pyruvate can be summarized as follows:

Glucose (Six carbon atoms) $\rightarrow$ 2 pyruvate (Three carbon atoms)

Summary of the reactions in glycolysis is given below.

Step 1. Phosphorylation of glucose to give glucose-6-phosphate (ATP is the source of the phosphate group).

Glucose + ATP → glucose-6-phosphate + ADP

Step 2. Isomerization of glucose-6-phosphate to give fructose-6-phosphate.

Glucose-6-phosphate → fructose-6-phosphate

Step 3. Phosphorylation of fructose-6-phosphate to give fructose-1,6-bisphosphate (ATP is the source of the phosphate group).

Fructose-6-phosphate + ATP →
fructose-1-6-bisphosphate + ADP

Step 4. Cleavage of fructose-1,6-bisphosphate to give two 3-carbon fragments, glyceraldehyde-3-phosphate and dihydroxyacetone phosphate.

Fructose-1,6-bisphosphate → glyceraldehyde-3-phosphate
+ dihydroxyacetone phosphate

Step 5. Isomerization of dihydroxyacetone phosphate to give glyceraldehyde-3-phosphate.

Dihydroxyacetone phosphate →
glyceraidehyde-3-phosphate

Step 6. Oxidation (and phosphorylation) of glyceraldehyde-3-phosphate to give 1,3-bisphosphoglycerate.

Glyceraldehyde-3-phosphate + NAD^+ + P → NADH +

1,3-bisphosphoglycerate + H+

Step 7. Transfer of a phosphate group from 1,3-UThosphoglycerate to ADP (phosphorylation of ADP to ATP) to give 3-phosphoglycerate.

1,3-Bisphosphoglycerate + ADP → 3-phosphoglycerate + ATP

Step 8. Isomerization of 3-phosphoglycerate to give 2-phosphoglycerate.

3-Phosphoglycerate → 2-phosphoglycerate

Step 9. Dehydration of 2-phosphoglycerate to give phosphoenolpyruvate.

2-Phosphoglycerate → phosphoenolpyruvate + H_2O

Step 10. Transfer of a phosphate group from phosphoenolpyruvate to ADP (phosphorylation of ADP to ATP) to give pyruvate.

Phosphoenolpyruvate + ADP → pyruvate + ATP

Conversion of glucose to glyceraldehyde-3-phosphate

The first steps of the glycolytic pathway prepare for the electron transfer and the eventual phosphorylation of ADP; these reactions make use of the free energy of hydrolysis of ATP.

Step 1. Glucose is phosphorylated to give glucose-6-phosphate. The phosphorylation of glucose is an endergonic reaction.

Glucose + Pi → glucose-6-phosphate + H_2O $\Delta G°' = 13.8$ kJ mol^{-1} = 3.3 kcal mol^{-1}

The hydrolysis of ATP is exergonic.

$$ATP + H_2O \rightarrow ADP + Pi \quad \Delta G^{\circ\prime} = -30.5 \text{ kJ mol}^{-1} = -7.3 \text{ kcal mol}^{-1}$$

These two reactions are coupled, so the overall reaction is the sum of the two and is exergonic.

$$\text{Glucose} + ATP \rightarrow \text{glucose-6-phosphate} + ADP \quad \Delta G^{\circ\prime} = 13.8 + (-30.5) = -16.7 \text{ kJ mol}^{-1} = -4.0 \text{ kcal mol}^{-1}$$

This reaction illustrates the use of chemical energy originally produced by the oxidation of nutrients and trapped by phosphorylation of ADP to ATP. ATP does not represent stored energy just as an electric current does not represent stored energy.

The chemical energy of nutrients is released by oxidation and is made available for immediate use on demand by being trapped as ATP. The enzyme that catalyzes this reaction is hexokinase. The term "kinase" is applied to the class of ATP-dependent enzymes that transfer a phosphate group from ATP to a substrate.

The substrate of hexokinase is not necessarily glucose, but it can be any one of a number of hexoses such as glucose, fructose, and mannose. Glucose-6-phosphate inhibits the activity of hexokinase; this is a control point in the pathway.

A large conformational change takes place in hexokinase when substrate is bound. It has been shown by x-ray crystallography that, in the absence of substrate, two lobes of the enzyme that surround the binding site are quite far apart.

When glucose is bound, the two lobes move closer together and the glucose becomes almost completely surrounded by protein. This type of behaviour is consistent with the induced-fit theory of enzyme action. In all kinases for which the structure is known, there is a cleft that closes when substrate is bound.

Step 2. Glucose-6-phosphate isomerizes to give fructose-6-phosphate. Glucosephosphate isomerase is the enzyme that catalyzes this reaction. The C-1 aldehyde group of glucose-6-phosphate is reduced to a hydroxyl group, and the C-2 hydroxyl group is oxidized to give the ketone group of fructose-6-phosphate, with no net oxidation or reduction. The phosphorylated forms, glucose-6-phosphate and fructose-6-phosphate, are an aldose and a ketose, respectively.

Step 3. Fructose-6-phosphate is further phosphorylated, producing fructose1,6-bisphosphate. The endergonic reaction of phosphorylation of fructose-6phosphate is coupled to the exergonic reaction of hydrolysis of ATP, and the overall reaction is exergonic ($\Delta G°' = -13.8 \text{ kJ mol}^{-1} = -3.3 \text{ kcal mol}^{-1}$).

$$\text{Fructose-6-phosphate} + \text{ATP} \rightarrow \text{fructose-1,6-bisphosphate} + \text{ADP}$$

The reaction in which fructose-6-phosphate is phosphorylated to give fructose-1,6-bisphosphate is the one in which the sugar is committed to glycolysis. Glucose-6--phosphate and fructose-6-phosphate can play roles in other pathways, but fructose-1,6-bisphosphate does not. Once fructose-1,6-bisphosphate is formed from the original sugar, no other pathways are available and the molecule must undergo the rest of the reactions of glycolysis. The phosphorylation of fructose-6-phosphate is highly exergonic and irreversible, and phosphofructokinase, the enzyme that catalyzes it, is the key regulatory enzyme in glycolysis.

Phosphofructokinase is a tetramer that is subject to

allosteric feedback regulation. This enzyme exists in slightly different forms in muscle and in liver; the two forms, designated M and L respectively, are referred to as isozymes. The subunits differ slightly in amino acid composition, so the two isozymes can be separated from each other by electrophoresis.

The tetrameric form that occurs in muscle is designated M4, whereas that in liver is designated L_4. In red blood cells, both isozymes, the M and the L forms, occur.

All possible combinations of monomers combine to form the tetramer: M_4, M_3L, M_2L_2, ML_3, and L_4. Individuals who lack the gene that directs the synthesis of the M form of the enzyme can carry on glycolysis in their livers but suffer muscle weakness because they lack the enzyme in muscle.

When the rate of the phosphofructokinase reaction is observed at varying concentrations of substrate (ftuctose-6-phosphate), the sigmoidal curve typical of allosteric enzymes is obtained. ATP is an allosteric effector in the reaction. High levels of ATP depress the rate of the reaction, and low levels of ATP stimulate the reaction.

When there is a high level of ATP in the cell, a good deal of chemical energy immediately available is from hydrolysis of ATP. The cell does not need to metabolize glucose for energy, so the presence of ATP inhibits the glycolytic pathway at this point. There is also another, more potent, allosteric effector of phosphofructokinase.

Step 4. Fructose-1,6-bisphosphate is then split into two 3-carbon fragments.

Fructose-1,6-bisphosphate $\rightarrow$ glyceraldehyde-3-phosphate + dihydroxyacetone phosphate

The cleavage reaction here is the reverse of an aldol condensation; the enzyme that catalyzes it is called aldolase. In the enzyme isolated from most animals sources (the one from muscle is the most extensively studied), the basic side chain of an essential lysine residue plays the key role in catalyzing this reaction. The thiol group of a cysteine also acts as a base here.

Step 5. The dihydroxyacetone phosphate is converted to glyceraldehyde-3phosphate.

The enzyme that catalyzes this reaction is triosephosphate isomerase. One molecule of glyceraldehyde-3-phosphate has already been produced by the aldolase reaction; we now have a second molecule of glyceraldehyde-3-phosphate, produced by the triosephosphate isomerase reaction. The original molecule of glucose, which contains six carbon atoms, has now been converted to two molecules of glyceraldehyde-3-phosphate, each of which contains three carbon atoms.

Conversion of glyceraldehyde-3-phosphate to pyruate

At this point a molecule of glucose that enters the pathway has been converted to two molecules of glyceraldehyde-3-phosphate. We have not seen any oxidation reactions yet, but now we shall encounter them. Keep in mind that in the rest of the pathway two molecules of each of the three-carbon compounds take part in every reaction for each original glucose molecule.

Step 6. The next step is the oxidation of glyceraldehyde-3-phosphate to 1, 3-bisphospho-glycerate. This reaction, the characteristic reaction of glycolysis, should be looked at more closely. It involves the addition of a phosphate group to glyceraldehyde-3-phosphate as well as an electron-transfer reaction, from glyceraldehyde-3-phosphate to NAD. It will simplify discussion to consider the two parts separately.

The half reaction of oxidation is that of an aldehyde to a carboxylic acid group, where water can be considered to take part in the reaction.

$$RCHO + H_2O \rightarrow RCOOH + 2H^+ + 2e^-$$

The half reaction of reduction is that of NAD+ to NADH.

$$NAD^+ + 2\ H^+ + 2C \rightarrow NADH + H^+$$

The overall redox reaction is thus

$$RCHO + H_2O + NAD^+ \rightarrow RCOOH + H^+ + NADH$$

Glyceraldehyde-3-phosphate 3-phosphoglyce rate

where R indicates the portions of the molecule other than the aldehyde and carboxylic acid groups, respectively. The oxidation reaction is exergonic under standard conditions ($\Delta G^{\circ\prime}$ = -43.1 kj mol^{-1} = -10.3 kcal mol-[1]), but oxidation is only part of the overall reaction.

The phosphate group that is linked to the carboxyl group does not form an ester, since an ester linkage requires an alcohol and an acid. Instead, the carboxylic acid group and phosphoric acid form a mixed anhydride of two acids by loss of water.

$$\text{3-Phosphoglycerate} + Pi \rightarrow \text{1,3-bisphosphoglycerate} + H_2$$

Where the substances involved in the reaction are in the ionized form appropriate at pH7. Note that ATP and ADP do not appear in the equation. The source of the phosphate group is phosphate ion itself, rather than ATP.

The phosphorylation reaction is endergonic under standard conditions ($\Delta G^{\circ\prime}$ 49.3 kj mol^{-1} = 11.8 kcal mol^{-1}).

The overall reaction, including electron transfer and phosphorylation, is

Glyceraldehyde-3-phosphate + Pi + $NAD^+ \rightarrow$ 1, 3-bis-phosphoglycerate + NADH + H^+

The standard free energy change for the overall reaction is the sum of the values for the oxidation and phosphorylation reactions. The overall reaction is not far from equilibrium, being only slightly endergonic.

$$\Delta G^{\circ\prime} \text{ overall} = \Delta G^{\circ\prime} \text{ oxidation} + \Delta G^{\circ\prime} \text{ Phosphorylation} = (-43.1) + (49.3) \text{ kj mol}^{-1} = 6.2 \text{ kj mol}^{-1} = 1.5 \text{ kcal mol}^{-1}$$

This value of the standard free energy change is for the reaction of one mole of glyceraldehyde-3-phosphate; the value must be multiplied by two to get the value for each mole of glucose ($\Delta G^{\circ\prime}$ = 12.4 kj = 3.0 kcal).

The enzyme that catalyzes the conversion of glyceraldehyde-3-phosphate to 1,3-bisphosphoglycerate is glyceraldehyde-3-phosphate dehydrogenase. This enzyme is one of a class of similar enzymes, the NADH-linked dehydrogenases.

The structures of a number of dehydrogenases of this type have been studied using x-ray crystallography. The overall structures are not strikingly similar, but the structure of the binding site for NADH is quite similar in all these enzymes. One portion of the binding site is specific for the nicotinamide ring, and one portion is specific for the adenine ring.

The molecule of glyceraldehyde-3-phosphate dehydrogenase is a tetramer, consisting of four identical subunits. Each subunit binds one molecule of NAD^+, and each subunit contains an essential cysteine residue. A thioester involving the cysteine residue is the key intermediate in this reaction.

In the phosphorylation step, the thioester acts as a "high-energy" intermediate. Phosphate ion attacks the thioester, forming a mixed anhydride of the carboxylic and phosphoric acids, which is also a "high-energy" compound. This compound is 1,3-bisphosphoglycerate, the product of the reaction. Production of ATP requires a "high-energy" compound as starting material. The 1,3-bisphosphoglycerate fulfills this requirement and transfers a phosphate group of ADP in a highly exergonic reaction.

Step 7. The next step is one of the two reactions in which ATP is produced by phosphorylation of ADP:

1, 3-bisphosphoglycerate + ADP ⇔ 1,3-phosphoglycerate + ATP

The enzyme that catalyzes this reaction is phosphoglycerate kinase. By now the term "kinase" should be familiar as the generic name for a class of ATP-dependent phosphate-group transfer enzymes. The most striking feature of the reaction has to do with energetics of phosphate-group transfer.

In this step in glycolysis, a phosphate group is transferred from 1,3, phosphoglycerate to a molecule of ADP, producing ATP, the first of two such reactions in the glycolytic pathway. We already mentioned that 1,3bisphosphoglycerate can easily transfer a phosphate group to other substances. Note that a substrate, namely 1.3-bisphosphoglycerate, has transferred a phosphate group to ADP. This transfer is typical of substrate-level phosphorylation.

It is to be distinguished from oxidative phosphorylation, in which transfer of phosphate groups is linked to electron-transfer reactions in which oxygen is the ultimate electron acceptor. The only requirement for substrate-level phosphorylation is that the standard free energy of the hydrolysis reaction is more negative than that for hydrolysis of the new phosphate compound being formed.

The standard free energy of hydrolysis of 1,3-bisphosphoglycerate is -49.3 kj mol^{-1}. We have already seen that the standard free energy of hydrolysis of ATP is -30.5 kj mol^{-1}, and we must change the sign of the free energy change when the reverse reaction occurs:

$$ADP + Pi + H^{+} \rightarrow ATP + H_2O,\ \Delta G^{\circ\prime} = 30.5 \text{ kj mol}^{-1} = 7.3 \text{ kcal mol}^{-1}$$

The net reaction is

$$\text{1,3-Bisphosphoglycerate} + ADP \rightarrow \text{3-phosphoglycerate} + ATP,\ \Delta G^{\circ\prime} = 48.8 \text{ kj mol}^{-1} = -4.5 \text{ kcal mol}^{-1}$$

Two molecules of ATP are produced by this reaction for each molecule of glucose that enters the glycolytic pathway. In the earlier stages of the pathway two molecules of ATP were invested to produce fructose-1,6-bisphosphate, and now they have been recovered. At this point the balance of ATP use and production is exactly even. The next few reactions will bring about the production of two more molecules of ATP for each original molecule of glucose, leading eventually to the net gain of two ATP molecules in glycolysis.

Step 8. In the next reaction the phosphate group is transferred from carbon 3 to carbon 2 of the glyceric acid backbone, setting the stage for the reactions that follow.

$$\text{3-phosphoglycerate} \Leftrightarrow \text{2-phosphoglycerate}$$

The enzyme that catalyzes this reaction is phosphoglyceromutase.

Step 9. The 2-phosphoglycerate molecule loses one molecule of water, producing phosphoenolpyruvate. This reaction does not involve electron transfer; it is a dehydration reaction. Enolase, the enzyme that catalyzes this reaction, requires Mg^{2+} as a cofactor. The water molecule that is eliminated binds to Mg^{2+} in the course of the reaction.

$$\text{2-Phosphoglycerate} \Leftrightarrow \text{phosphoenolpyruvate} + H_2O$$

Step 10. Phosphoenolpyruvate then transfers its phosphate group to ADP, producing ATP and pyruvate.

$$\text{Phosphoenolpyruvate} + \text{ADP} \rightarrow \text{pyruvate} + \text{ATP}$$

The double bond shifts to the oxygen on carbon 2 and a hydrogen shifts to carbon 3. Phosphoenolpyruvate is a "high-energy" compound with a high phosphate-group transfer potential. The free energy of hydrolysis of this compound is more negative than that of ATP.

The reaction that occurs in this step can be considered to be the sum of the hydrolysis of phosphoenolpyruvate and the phosphorylation of ADR This reaction is another example of substrate-level phosphorylation.

$$\text{Phosphoenolpyruvate} \rightarrow \text{pyruvate} + \text{Pi} \ \Delta G°' = -61.9 \text{ V mol}^{-1} = -14.8 \text{ kcal mol-}^{1}$$

$$\text{ADP} + \text{Pi} \rightarrow \text{ATP} \ \Delta G°' = 30.5 \text{ kJ mol}^{-1} = 7.3 \text{ kcal mol}^{-1}$$

The net reaction is

$$\text{Phosphoenolpyruvate} + \text{ADP} \rightarrow 6 \text{ pyruvate} + \text{ATP} \ \Delta G°'$$

$$= -31.4\text{-kj mol}^{-1} = -7.5 \text{ kcal mol}^{-1}$$

Since two moles of pyruvate are produced for each mole of glucose, twice as much energy is released for each mole of starting material. Pyruvate kinase is the enzyme that catalyzes this reaction. Like phosphofructokinase, it is an allosteric enzyme consisting of four subunits.

Pyruvate kinase is inhibited by ATP. The conversion of phosphoenolpyruvate to pyruvate slows down when the cell has a high concentration of ATP, that is to say, when the cell does not have a great need for energy in the form of ATP.

Control points

One of the most important questions that we can ask about any metabolic pathway is that of the points at which control is exercised over it. Pathways can be "shut down" if any organism has no immediate need for their products, saving energy for the organism. In glycolysis, three reactions are control points: the first reaction of glucose to glucose-6-phosphate, catalyzed by hexokinase; the production of fructose-1,6-bisphosphate, catalyzed by phospho-fructokinase; and the last reaction of the pathway, catalyzed by pyruvate kinase

It is frequently observed that control is exercised near the start and end of a pathway, as well as at points involving key intermediates such as fructose-1,6-bisphosphate. When we have learned more about the metabolism of carbohydrates, we can return to the role of phosphofructokinase and fructose-1,6-bisphosphate in the regulation of several pathways of carbohydrate metabolism.

Anaerobic reaction of pyruates

The final reaction of anaerobic glycolysis is the reduction

of pyruvate to lactate. This reaction is also exergonic ($\Delta G^{\circ\prime}$ = -25.1 kJ mol^{-1} = -6.0 kcal mol^{-1}); as before, we need to multiply this value by two to find the energy yield for each molecule of glucose that enters the pathway.

Lactate is a dead end in metabolism, but it can be recycled to form pyruvate and even glucose by a pathway called gluconeogenesis. Lactate dehydrogenase (LDH) is the enzyme that catalyzes this reaction. Like glyceraldehyde-3-phosphate dehydro-genase, LDH is an NADH-linked dehydrogenase and consists of four subunits. There are two kinds of subunits, designated M and H, which vary slightly in amino acid composition.

The quaternary structure of the tetramer can vary according to the relative amounts of the two kinds of subunits. These variant forms of the enzyme are called isozymes. In human skeletal muscle the homogeneous tetramer of the M_4 type predominates, and in the heart the other homogeneous possibility, the H_4 tetramer, is the predominant form. The heterogeneous forms, M_3H, M_2H_2, and MH_3, occur in blood serum.

A very sensitive clinical test for heart disease is based on the existence of the various isozymic forms of this enzyme. The relative amounts of the H_4 and MH_3 isozymes in blood serum increase drastically after myocardial infarction (heart attack) compared with normal serum. At this point one might ask why the reduction of pyruvate to lactate (a waste product in-aerobic organisms) is the last step in anaerobic glycolysis, a pathway that provides energy for the organism by oxidation of nutrients.

There is another point to consider about the reaction, one that involves the relative amounts of NAD^+ and NADH in a cell. The half reaction of reduction can be written

$$\text{Pyruvate} + 2H^{+} + 2e^{-} \rightarrow \text{lactate}$$

The half reaction of oxidation is

$$\text{NADH} + H^{+} \rightarrow \text{NAD}^{+} + 2C + 2H^{+}$$

The overall reaction is,

$$\text{Pyruvate} + \text{NADH} + \text{HI} \rightarrow \text{lactate} + \text{NAD}^{+}$$

The NADH produced from NAD^+ by the earlier oxidation of glyceraldehyde-3-phosphate is used up with no net change in the relative amounts of NADH and NAD^+ in the cell. This regeneration is needed under anaerobic conditions in the cell so that NAD^+ will be present for further glycolysis to take place.

Without this regeneration, the oxidation reactions in anaerobic organisms would soon come to a halt because of the lack of NAD^+ to serve as an oxidizing agent in fermentative processes. On the other hand, NADH is a frequently encountered reducing agent in many reactions, and it is lost to the organism in lactate production.

Aerobic metabolism makes more efficient use of reducing agents ("reducing power") such as NADH, because the conversion of pyruvate to lactate does not occur in aerobic metabolism.

The NADH produced in the stages of glycolysis leading to the production of pyruvate is available for use in other reactions in which a reducing agent is needed.

Anaerobic metabolism and tooth decay

Dental caries, tooth decay, is one of the most prevalent diseases in the United States and possibly in the world, although modern treatments such as fluoride and flossing have greatly reduced its incidence in young people.

Contributing factors in tooth decay are a combination of a diet high in refined sugars, the development of dental plaque, and anaerobic metabolism. The high-sugar diet allows for ready growth of bacteria in the mouth, and sucrose is perhaps the most efficiently used sugar, since the bacteria can make their polysaccharide "glue" more efficiently from this nonreducing sugar.

The bacteria grow in expanding sticky colonies, forming plaque on the tooth surface. The bacteria growing under the surface of the plaque have anaerobic metabolism, since oxygen does not diffuse readily through the waxy surface of dental plaque.

The two predominant by-products, lactate and pyruvate, are relatively strong organic acids, and these acid products are what actually cause destruction of the enamel surface. The bacteria, of course, grow rapidly in the pock holes and if the enamel is eaten all the way through, the bacteria grow even more readily in the softer dentin layer beneath the enamel. Fluoridation results in a much harder enamel surface, and fluoride may actually inhibit metabolism.

Alcoholic fermentation

Two reactions related to the glycolytic pathway lead to the production of ethanol by alcoholic fermentation. This process is one of the alternative fates of pyruvate. In the first of the two reactions that lead to the production of ethanol, pyruvate is decarboxylated (loses carbon dioxide) to produce acetaldehyde. The enzyme that catalyzes this reaction is pyruvate decarboxylase. This enzyme requires Mg^{2+} and a cofactor we have not met before, thiamine pyrophosphate (TPP). (Thiamine itself is vitamin III.)

In TPP the carbon atom between the nitrogen and

the sulphur in the thiazole ring is highly reactive. It forms a carbanion (an ion with a negative charge on a carbon atom) quite easily, and the carbanion in turn attacks the carbonyl group of pyruvate to form an adduct. Carbon dioxide splits off, leaving a two-carbon fragment covalently bonded to TPP.

There is a shift of electrons, and the two-carbon fragment splits off, producing acetaldehyde. (The two-carbon fragment bonded to TPP is sometimes called activated acetaldehyde.)

$$\text{pyruvate} \rightarrow \text{acetaldehyde} + CO_2$$

The carbon dioxide produced is responsible for the bubbles in beer and sparkling wines. Acetaldehyde is then reduced to produce ethanol, and at the same time one molecule of NADH is oxidized to NAD+ for each molecule of ethanol produced.

$$\text{Acetaldehyde} + \text{NADH} \rightarrow \text{ethanol} + \text{NAD}^+$$

The reduction reaction of alcoholic fermentation is similar to the reduction of pyruvate to lactate, in the sense that it provides for recycling of NAD+ and thus allows further anaerobic oxidation (fermentation) reactions. The net reaction for alcoholic fermentation is

$$\text{Glucose} + 2\ \text{ADP} + 2\ \text{Pi} + 2H^+ \rightarrow 2\ \text{ethanol} + 2\text{ATP} + 2\ CO_2 + 2\ H_2O$$

NAD^+ and NADH do not appear explicitly in the net equation. It is essential that the recycling of NADH to NAD^+ takes place here, just as it does when lactate is produced, so that there can be further anaerobic oxidation. Alcohol dehydrogenase, the enzyme that catalyzes the conversion of acetaldehyde to ethanol, is similar to lactate dehydrogenase in many ways.

The most striking similarity is that both are NADH-linked dehydrogenases, and both are tetramers. The complex of injuries to a fetus caused by maternal consumption of ethanol is called fetal alcohol syndrome. In catabolism of ethanol by the body, the first step is conversion to acetaldehyde, the reverse of the last reaction of alcoholic fermentation. The level of acetaldehyde in the blood of a pregnant woman is the key to detecting fetal alcohol syndrome.

It has recently been shown that the acetaldehyde is transferred across the placenta and accumulates in the liver of the fetus. The acetaldehyde in turn is easily oxidized to breakdown products and is responsible for the harmful effects. The labels of alcoholic beverages now include a warning against consumption during pregnancy.

Chapter 5

Citric Acid Cycle

In the first stage of energy extraction from carbohydrates, glucose is catabolized to pyruvate in the ten anaerobic steps of glycolysis. In aerobic catabolism, pyruvate is converted to "high-energy" two-carbon acetyl-CoA, which then enters the citric acid cycle. As the first step of the cycle, acetyl-CoA combines with oxaloacetate to form citric acid. Each turn of the cycle releases two molecules of CO_2 regenerates the starting molecule oxaloacetate, and delivers reducing agents to the electron transport chain. In the final stage of glucose metabolism, proton flow across the inner mitochondrial membrane creates ATP, the energy currency of the cell.

The citric acid cycle not only furnishes energy, it also provides intermediates for the biosynthesis of proteins, lipids, and the heme group. The citric acid cycle, as the hub of metabolic pathways, serves to connect the breakdown and synthesis of proteins, carbohydrates, and lipids. Most of the metabolites of major nutrients can be fed into the citric acid cycle as acetyl-CoA and oxidized to produce energy. Alternatively, intermediates of the citric acid cycle can be converted to other biomolecules.

The evolution of aerobic metabolism, by which nutrients are oxidized to carbon dioxide and water, was an important step in the history of life on Earth.

Organisms can obtain far more energy from nutrients by aerobic oxidation than by anaerobic oxidation. Metabolism consists of catabolism, which is oxidative breakdown of nutrients, and anabolism, which is reductive synthesis of biomolecules.

The citric acid cycle is amphibolic, meaning that it plays a role in both catabolism and anabolism. Although the citric acid cycle is a part of the pathway of aerobic oxidation of nutrients, some of the molecules that are included in this cycle are the starting points of biosynthetic (anabolic) pathways. Metabolic pathways operate simultaneously, even though we talk about them separately. We should always keep this point in mind. There are two other common names for the citric acid cycle. One is the Krebs cycle, after Sir Hans Krebs, who first investigated the pathway. The other name is the tricarboxylic acid cycle, from the fact that some of the molecules involved are acids with three carboxyl groups.

An important difference between glycolysis and the citric acid cycle is the part of the cell in which these pathways occur. In prokaryotes glycolysis takes place in the cytosol, whereas the citric acid cycle takes place on the plasma membrane.

In eukaryotes glycolysis occurs in the cytosol, whereas the citric acid cycle takes place in mitochondria. Most of the enzymes of the citric acid cycle are present in the mitochondrial matrix. A quick review of some aspects of mitochondrial structure is in order here, since we shall want to describe the exact location of each of the components of the citric acid cycle and the electron transport chain.

A mitochondrion has an inner and an outer membrane. The region enclosed by the inner membrane is called the matrix, and there is an intermembrane space between the inner and outer membranes. The reactions of the citric acid cycle take place in the matrix, except for the one in which the intermediate electron acceptor is FAD (flavin adenine dinucleotide). The enzyme that catalyzes the FAD-linked reaction is an integral part of the inner mitochondrial membrane.

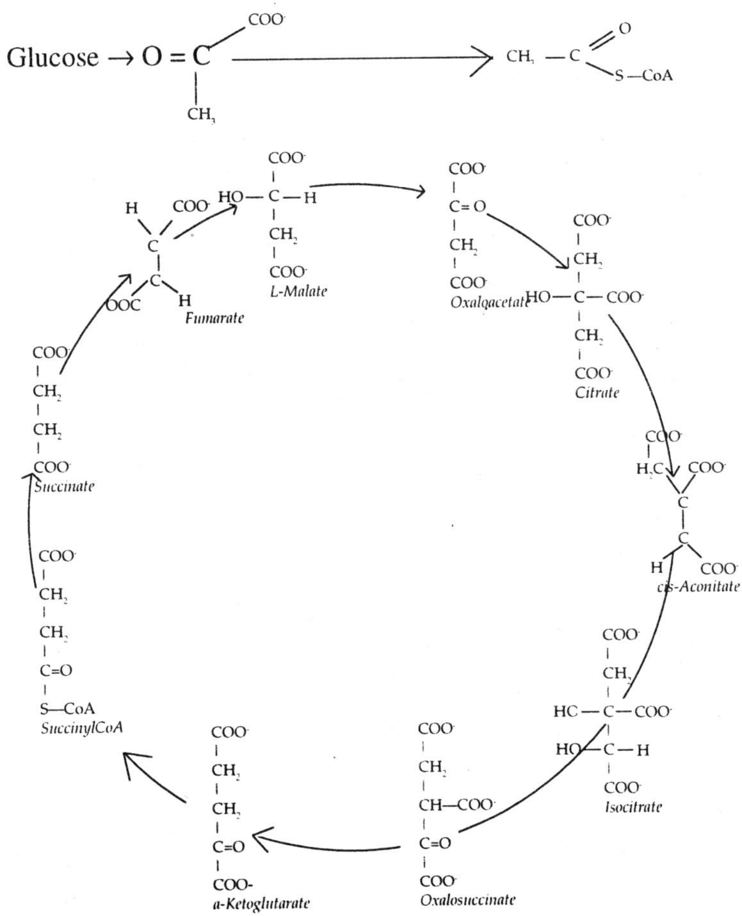

Fig.1. Cytric acid cycle

The citric acid cycle is amphibolic. It plays a role in both catabolism and anabolism. It is the central metabolic pathway. The citric acid cycle is shown in schematic form in Figure 1.

Under aerobic conditions, pyruvate produced by glycolysis is oxidized further, with carbon dioxide and water as the final products. First the pyruvate is oxidized to one carbon dioxide molecule and to one, acetyl group, which becomes linked to an intermediate, coenzyme A (CoA). The acetyl-CoA enters the citric acid cycle. In the citric acid cycle two more molecules of carbon dioxide are produced for each molecule of acetyl-CoA that enters the cycle, and electrons are lost in the process. The immediate electron acceptor in all cases but one is NAD^{+}.

In the one case where there is another intermediate electron acceptor, it is FAD, which takes up electrons and hydrogen to produce $FADH_2$, the reduced form of flavin derived from riboflavin. The electrons are passed from NADH and $FADH_2$ through several stages of an electron transport chain with a different redox reaction at each step. The final electron acceptor is oxygen, with water as the product.

In the first reaction of the cycle the two-carbon acetyl group condenses with the four-carbon oxaloacetate ion to produce the six-carbon citrate ion. In the next few steps the citrate isomerizes, and then it both loses carbon dioxide and is oxidized. This process, called oxidative decarboxylation, produces the five-carbon compound a-ketoglutarate, which again is oxidatively decarboxylated to produce the four-carbon compound succinate. The cycle is completed by regeneration of oxaloacetate from succinate in several steps. There are eight steps in the citric acid cycle, each catalyzed by a different enzyme. Four of the eight steps are oxidation reactions: those in steps 3, 4, 6, and 8. The oxidizing agent is NAD' in all except Step 6, in which FAD plays the same role. In step 5 a molecule of GDP (guanosine diphosphate) is phosphorylated to produce GTP (guanosine triphosphate). This reaction is equivalent to the production of ATP, since the phosphate group is easily transferred to ADP, producing GDP and ATP.

Reactions in citric acid cycle

Conversion of pyruvate to acetyl-CoA

An enzyme system called the pyruvate dehydrogenase complex is responsible for the conversion of pyruvate to carbon dioxide and the acetyl portion of acetyl-CoA. There is an -SH group at one end of the CoA molecule, which is the point at which the acetyl group is attached. As a result, CoA is frequently shown in equations as CoA-SH. Because CoA is a thiol (the sulphur [thiol analog of an alcohol), acetyl-CoA is a thioester, with a sulphur atom replacing an oxygen of the usual carboxylic ester. This difference is important, since thioesters are "high-energy" compounds. In other words, the hydrolysis of thioesters releases enough energy to drive other reactions. An oxidation reaction precedes the transfer of the acetyl group to the CoA. The whole process involves several enzymes, all of which are part of the pyruvate dehydrogenase complex.

The overall reaction

$$\text{Pyruvate} + \text{CoA-SH} + \text{NAD}^+ \rightarrow \text{acetyl-CoA} + CO_2 + H^+ + \text{NADH}$$

The reaction is exergonic ($\Delta G^{\circ\prime} = -33.4 \text{ kj mol}^{-1} = -8.0 \text{ kcal mol}^{-1}$), and NADH is recycled for further use.

Three enzymes make up the pyruvate dehydrogenase complex. They are pyruvate dehydrogenase, dihydrolipoyl transacetylase, and dihydrolipoyl dehydrogenase. The reaction takes place in four steps. The last two enzymes catalyze reactions of lipoic acid, a compound that has a disulphide group in its oxidized form and two sulfhydryl groups in its reduced form.

Lipoic acid differs in one respect from other coenzymes. It is a vitamin, rather than a metabolite of a vitamin, as is the case with many other coenzymes. Lipoic acid can act as an oxidizing agent; the reaction involves hydrogen transfer, which frequently accompanies biological oxidation-reduction reactions.

Another reaction of lipoic acid is the formation of a thioester linkage with the acetyl group before it is transferred to the acetyl-CoA. Lipoic acid can act simply as an oxidizing agent, or it call simultaneously take part in two reactions, a redox reaction and the shift of an acetyl group by transesterification. The first step in the reaction sequence that converts pyruvate to carbon dioxide and acetyl-CoA is catalyzed by pyruvate dehydrogenase. This enzyme requires thiamine pyrophosphate as a coenzyme. The coenzyme is not covalently bonded to the enzyme; they are held together by noncovalent interactions. Mg^{2+} is also required. We saw the action of TPP as a coenzyme in the conversion of pyruvate to acetaldehyde, catalyzed by pyruvate decarboxylase.

In the pyruvate dehydrogenase reaction an a-keto acid, pyruvate, loses carbon dioxide; the examining two-carbon unit becomes covalently bonded to TPP. The second step of the reaction is catalyzed by dihydrolipoyl transacetylase. This enzyme requires lipoic acid as a coenzyme. The lipoic acid is covalently bonded to the enzyme by an amide bond to the ε-amino group of a lysine side chain.

The two-carbon unit that originally came from pyruvate is transferred from the thiamine pyrophosphate to the lipoic acid and in the process a hydroxyl group is oxidized to produce an acetyl group. The disulphide group of the lipoic acid is the oxidizing agent, which is itself reduced, and the product of the reaction is a thioester (an ester in which an oxygen is replaced by a sulphur). In other words, the acetyl group is now covalently bonded to the lipoic acid by a thioester linkage.

The third step of the reaction is also catalyzed by dihydrolipoyl transacetylase. A molecule of CoA-SH attacks the thioester linkage, and the acetyl group is transferred to it. The acetyl group remains bound in a thioester linkage, this time as acetyl-CoA rather than esterified to lipoic acid. Tile reduced form of lipoic acid remains covalently bound to dihydrolipoyl transacetylase. The reaction of pyruvate and CoA-SH has now reached the stage of the products, carbon dioxide and acetyl-CoA, but the lipoic acid coenzyme is in a reduced form. If the transacetylase is to catalyze further reactions, the lipoic acid must be regenerated.

In the fourth step of the overall reaction, the enzyme dihydrolipoyl dehydrogenase reoxidizes the reduced lipoic acid from the sulfhydryl to the disulphide form. The lipoic acid still remains covalently bonded to the transacetylase enzyme. The dehydrogenase also has a coenzyme, FAD, which is bound to the enzyme by noncovalent interactions. An electron acceptor is simply another name for an oxidizing agent, and the coenzyme serves this function here. As, a result, FAD is reduced to $EADH_2$. $FADH_2$ is reoxidized in turn. The oxidizing agent is NAD^+, and NADH is the product along with reoxidized FAD. The reduction of NAD^+ to NADH accompanies the oxidation of pyruvate to the acetyl group, and the overall equation shows that there has been a transfer of two electrons from pyruvate to NAD^+ .The electrons gained by NAD^+ in generating NADH -in this step are passed to the electron transport chain.

Two molecules of pyruvate are produced for each molecule of glucose, so that there will eventually be five ATP from each glucose from this step alone. The reaction leading from pyruvate to acetyl-CoA is a complex one that requires three enzymes, each of which has its own co-enzyme in addition to NAD^+. The spatial orientation of the individual enzyme molecules with respect to one another is itself complex. In the enzyme isolated from E. coli the arrangement is quite compact so that the various steps of the reaction can be thoroughly coordinated.

There is a core of 24 dihydrolipoyl transacetylase molecules. The 24 polypeptide chains are arranged in eight trimers, with each trimer occupying the corner of a cube. There are 24 ap dimers, of pyruvate dehydrogenase, and they occupy the edges of the cube. Finally, six dimers of dihydrolipoyl dehydrogenase lie on the six faces of the cube. Note that many levels of structure combine to produce a suitable environment for the conversion of pyruvate to acetyl-CoA. Each of the enzyme molecules in this array has its own tertiary structure, and the array itself has the cubical structure we have just seen.

A compact arrangement such as the one in the pyruvate dehydrogenase multienzyme complex has two great advantages over an arrangement in which the various. Components are more widely dispersed. First, the various stages of the reaction can take place more efficiently because the reactants and the enzymes are so close to each other. The role of lipoic acid is particularly important here. Recall that the lipoic acid is covalently attached to the transacetylase enzyme that occupies a central position in the complex.

The lipoic acid and the lysine side chain to which it is bonded are long enough to act as a "swinging arm," which can move to the site of each of the steps of the reaction. As a result of the swinging arm action, the lipoic acid can move to the pyruvate dehydrogenase site to accept the two-carbon unit and then transfer it to the active site of the transacetylase. The acetyl group can then by transesterified to CoA-SH from the lipoic acid. Finally, the lipoic acid can swing to the active site of the dehydrogenase so that the sulfhydryl groups can be reoxidized to a disulphide.

A second advantage of a multienzyme complex is that regulatory controls can be applied more efficiently in such a system than with a single enzyme molecule. In the case of the pyruvate dehydrogenase complex, controlling factors are intimately associated with the multienzyme complex itself.

The overall reaction is part of a pathway that releases energy. It is not surprising that the enzyme that catalyzes it is inhibited by ATP and NADH, since both compounds are abundant when a cell has a good deal of energy readily available. The end products of a series of reactions inhibit the first reaction of the series, and the intermediate reactions do not take place when their products are not needed. It is consistent with this picture that the pyruvate dehydrogenase complex is activated by ADP, which is abundant when a cell needs energy. In mammals the actual mechanism by which the inhibition takes place is the phosphorylation of pyruvate dehydrogenase. A phosphate group is covalently bound to the enzyme in a reaction catalyzed by the enzyme pyruvate dehydrogenase kinase.

When the need arises for pyruvate dehydrogenase to be activated, the hydrolysis of the phosphate ester linkage (dephosphorylation) is catalyzed by another enzyme, phosphoprotein phosphatase. This latter enzyme is itself activated by Ca^{2+} Both of these enzymes are associated with the intact pyruvate dehydrogenase complex, permitting effective control of the overall reaction from pyruvate to acetyl-CoA.

Citric acid cycle proper

Reactions of the citric acid cycle proper and the enzymes that catalyze them are discussed below.

Step 1. Formation of citrate

The first step of the citric acid cycle is the reaction of acetyl-CoA and oxaloacetate to form citrate and CoA-SH. This reaction is called a condensation because a new carbon-carbon bond is formed. The condensation reaction of acetyl-CoA and oxaloacetate to formcitryl-CoA takes place in the first stage of the reaction. The condensation is followed by the hydrolysis of citryl-CoA to give citrate and CoA-SH. The two-carbon unit needed at the start of the citric acid cycle is obtained by conversion of pyruvate to the acetyl portion of acetyl-CoA. This conversion requires the three enzymes of the pyruvate dehydrogenase complex, as well as the cofactors TPP, FAD, NAD^+ and lipoic acid.

The reaction is catalyzed by the enzyme citrate synthase, originally called condensing enzyme. It is an exergonic reaction because the hydrolysis of a thioester releases energy are considered "high-energy" compounds.Both ATP and NADH inhibit the citrate synthase reaction, as we saw earlier with the pyruvate dehydrogenase reaction. Succinyl-CoA, a substance that appears later in the cycle, is an inhibitor as well. This is another example of the type of feedback control mechanism in which products of a long series of reactions "turn off" the first reaction of the series.

Step 2. Isomerization of citrate to isocitrate

The second reaction of the citric acid cycle, the one catalyzed by aconitase, is the isomerization of citrate to isocitrate. The enzyme requires Fe^{2+}. One of the most interesting features of the reaction is that citrate, a symmetrical (achiral) compound, is converted to isocitrate, a chiral compound, a molecule that cannot be superimposed on its mirror image. It is often possible for a chiral compound to have several different isomers. Isocitrate has four possible isomers, but only one of the four is produced by this reaction. Aconitase, the enzyme that catalyzes the conversion of citrate to isocitrate, is able to select one end of the citrate molecule in preference to the other. Another possible substrate for citrate synthase is fluoroacetyl-CoA.

The source of the fluoroacetyl-CoA is fluoroacetate, which is found in the leaves of various types of poisonous plants, including locoweeds. Animals that ingest these plants form fluoroacetyl-CoA, which in turn is converted to fluorocitrate by their citrate synthase. Fluorocitrate in turn is a potent inhibitor of aconitase, the enzyme that catalyzes the next reaction of the citric acid cycle. These plants are poisonous because they produce a potent inhibitor of life processes. The poison "ten eighty" (1080) used by sheep ranchers is sodium fluoroacetate. Ranchers who want to protect their sheep from attacks by coyotes put the poison just outside the ranch fence.

When the coyotes eat this poison they die, The mechanism of poisoning by "teneighty" is the same as that by plant poisons. This type of behavior means that the enzyme can bind a symmetrical substrate in an unsymmetrical binding site. The enzyme forms an unsymmetrical three-point attachment to the citrate molecule. The reaction proceeds by removal of a water molecule from the citrate to produce cis-aconitate, and then water is added back to the cis-aconitate to give isocitrate. The intermediate, cis-aconitate, remains bound to the enzyme during the course of the reaction. There is some evidence that the citrate is complexed to the Fe(II) in the active site of the enzyme in such a way that the citrate curls back on itself in a nearly circular conformation. Several authors have been unable to resist the temptation to call this situation the "ferrous wheel."

Step 3. Formation of a-ketoglutarate and CO_2—firstoxidation

The third step in the citric acid cycle is the oxidative decarboxylation of isocitrate to α-ketoglutarate and carbon dioxide. This reaction is the first of two oxidative decarboxylations of the citric acid cycle; the enzyme that catalyzes it is isocitrate dehydrogenase. The reaction takes place in two steps. First, isocitrate is oxidized to oxalosuccinate, which remains bound to the enzyme. Then oxalosuccinate is decarboxylated, and the carbon dioxide and a-ketoglutarate are released. This is the first of the reactions in which NADH is produced.

One molecule of NADH is produced from NAD^+ at this stage by the loss of two electrons in the oxidation. As we saw in our discussion of the pyruvate dehydrogenase complex, each NADH produced will lead to the production of 2.5 ATP in later stages of aerobic metabolism. Recall also that there will be two NADH, equivalent to five ATP for each original molecule of glucose. This reaction is a control site in the cycle. The enzyme is an oligomer, with a molecular weight on the order of 105. The reaction is controlled by the allosteric mechanism discussed in Chapter 5. The enzyme is inhibited by ATP and NADH and is activated by ADP and NAD^+ This pattern of inhibition by ATP and NADH and activation by ADP and NAD^+ is frequently encountered in catabolic reactions.

Step 4. Formation of succinyl-CoA and CO_2—econd oxidation

The second oxidative decarboxylation takes place in Step 4 of the citric acid cycle, in which carbon dioxide and succinyl-CoA are formed from α-ketoglutarate and CoA. This reaction is similar to the one in which acetyl-CcA is formed from pyruvate with NADH produced from NAD′. Once again each NADH will eventually give rise to 2.5 ATP, with five ATP from each original molecule of glucose. The reaction occurs in several stages and is catalyzed by an enzyme system called the α-ketoglutarate dehydrogenase complex, which is very similar to the pyruvate dehydrogenase complex.

Each of these multienzyme systems consists of three enzymes. The reaction takes place in several steps, and there is again a requirement for thiamine pyrophosphate (TPP), FAD, lipoic acid, and Mg^{2+}. This reaction is highly exergonic, as is the onecatalyzed by pyruvate dehydrogenase. It is also a control point in the pathway. At this point, two molecules of CO_2 have been produced by the oxidative decarboxylations of the citric acid cycle. Removal of the CO_2 makes the citric acid cycle irreversible in vivo, although in vitro each separate reaction is reversible. One might suspect that the two molecules of CO_2 arise from the two carbon atoms of acetyl-CoA.

Labelling studies have shown that this is not the case, but a full discussion of this point is beyond the scope of this text. The two CO_2 molecules arise from carbon atoms that were part of the oxaloacetate with which the acetyl group condensed. The carbons of this acetyl group are incorporated into the oxaloacetate that will be regenerated for the next round of the cycle. We should also mention that the α-ketoglutarate dehydrogenase complex reaction is the third one in which we have encountered an enzyme that requires

TPP. In all three bases the pyruvate dehydrogenase complex, the α-ketoglutaxate dehydrogenase complex, and pyruvate decarboxylase. The reaction involves transfer of an activated two-carbon group and the decarboxylation of an α–keto acid.

Step 5. Formation of succinate

In the next step of the cycle, the thioester bond of succinyl-CoA is hydrolyzed to produce succinate and CoA-SH; an accompanying reaction is the phosphorylation of GDP to GTP. The whole reaction is catalyzed by the enzyme succinyl-CoA synthetase. In the reaction mechanism, a phosphate group covalently bonded to the enzyme is directly transferred to the GDP.

The phosphorylation of GDP to GTP is endergonic, as is the corresponding ADP-ATP reaction. The energy required for the phosphorylation of GDP to GTP is provided by the hydrolysis of succinyl-CoA to produce succinate and CoA. The free energy of hydrolysis of succinyl-CoA is -33.4 kj mol $^{-1.}$ This reaction is similar to the one in which acetyl-CoA is formed from pyruvate, with NADH produced from NAD′. Once again each NADH will eventually give rise to 2.5 ATP, with five ATP from each original molecule of glucose. The reaction occurs in several stages and is catalyzed by an enzyme system called the α-ketoglutarate dehydrogenase complex, which is very similar to the pyruvate dehydrogenase complex.

Each of these multienzyme systems consists of three enzymes. The reaction takes place in several steps, and there is again a requirement for thiamine pyrophosphate (TPP), FAD, lipoic acid, and Mg^{2+}. This reaction is highly exergonic as is the one catalyzed by pyruvate dehydrogenase. It is also a control point in the pathway. At this point, two molecules of CO_2 have been produced by the oxidative decarboxylations of the citric acid cycle. Removal of the CO_2 makes the citric acid cycle irreversible in vivo, although in vitro each separate reaction is reversible. One might suspect that the two molecules of CO_2 arise from the two carbon atoms of acetyl-CoA.

Labelling studies have shown that this is not the case, but a full discussion of this point is beyond the scope of this text. The two CO_2 molecules arise from carbon atoms that were part of the oxaloacetate with which the acetyl group condensed. The carbons of this acetyl group are incorporated into the oxaloacetate that will be regenerated for the next round of the cycle. We should also mention that the a-ketoglutarate dehydrogenase complex reaction is the third one in which we have encountered an enzyme that requires TPP. In all three cases the pyruvate dehydrogenase complex, the α-ketoglutarate dehydrogenase complex, and pyruvate decarboxylase, the reaction involves transfer of an activated two-carbon group and the decarboxylation of an a-keto acid.

The overall reaction is slightly exergonic and, as a result, does not contribute greatly to the overall production of energy by the mitochondrion. The enzyme nucleoside diphosphate kinase catalyzes the transfer of a phosphate group from GTP to ADP to give GDP and ATP.

$$GTP + ADP \rightarrow GDP + ATP$$

This reaction step is called substrate-level phosphorylation to distinguish it from the type of reaction for production of ATP that is coupled to the electron transport chain.

The production of ATP in this reaction is the only

place in the citric acid cycle in which chemical energy in the form of ATP is made available to the cell. Except for this reaction, the generation of ATP characteristic of aerobic metabolism is associated with the electron transport chain, the subject of the next chapter. About 30 molecules of ATP (there is some question about the exact stoichiometry) can be obtained from the oxidation of a single molecule of glucose by the combination of anaerobic and aerobic oxidation, compared with only two molecules of ATP produced by anaerobic glycolysis alone. The combined reactions that occur in mitochondria are of great importance to aerobic organisms.

Step 6. Formation of fumarate—FAD-linked oxidation

In Step 6, succinate is oxidized to fumarate, a reaction that is catalyzed by the enzyme succinate dehydrogenase. This enzyme is an integral protein of the inner mitochondrial membrane. We shall have much more to say about the enzymes bound to the inner mitochondrial membrane in the next chapter. The other individual enzymes of the citric acid cycle are in the mitochondrial matrix. The electron acceptor, which is FAD rather than NAD^+, is covalently bonded to the enzyme; succinate dehydrogenase is called a flavoprotein because of the presence of FAD with its flavin moiety. In the succinate dehydrogenase reaction, FAD is reduced to $FADH_2$ and succinate is oxidized to fumarate.

$$\text{Succinate} + \text{E-FAD} \rightarrow \text{fumarate} + \text{E-FADH}_2$$

The E-FAD and E-$FADH_2$ in the equation indicate that the electron acceptor is covalently bonded to the enzyme. The $FADH_2$ group also passes electrons on to the electron transport chain and eventually to oxygen but will give rise to 1.5 ATP, rather than 2.5 ATP, as is the case with NADH. As a result, each original glucose will give rise to fewer ATP from this step. Succinate dehydrogenase contains iron atoms, but does not contain a heme group; it is referred to as a nonheme iron protein. The latter name refers to the fact that the protein contains several clusters that consist of four atoms each of iron and of sulphur. This enzyme exhibits stereospecificity.

Succinate itself is an achiral compound; when it is formed, the asymmetry of earlier compounds in the cycle is lost. The introduction of the double bond introduces another possibility for different isomers. The double bond produced by this reaction could have either the trans configuration (fumarate) or the cis configuration (maleate), but only the trans configuration is actually produced. This reaction is not a regulatory control point of the citric acid cycle.

Step 7. Formation oft-malate

In reaction 7, which is catalyzed by the enzyme fumarase, water is added across the double bond of fumarate in a hydration reaction to give malate. Again there is stereospecificity in the reaction. Malate has two enantiomers, L- and D-malate, but only L-malate is produced. This reaction is another one in which no control of the cycle is apparent. In the citric acid cycle and the pyruvate dehydrogenase reaction, one molecule of pyruvate is oxidized to three molecules of carbon dioxide. In the process, four NAP^+ are reduced to NADH, and one FAD to $FADH_2$; in addition one GDP is phosphorylated to GTP

Step 8. Regeneration of oxaloacetate —final oxidation step

In Step 8, malate is oxidized to oxaloacetate, and another molecule of NAD^+ is reduced to NADH. This reaction is catalyzed by the enzyme malate dehydrogenase and is not a control point for the cycle. This reaction is similar to that of other NADH-linked clehydrogenases, which we have seen previously. The oxaloacetate can then react with another molecule of acetyl-CoA to start another round of the cycle.

The oxidation of pyruvate by the pyruvate dehydrogenase complex and the citric acid cycle results in the production of three molecules Of CO_2 . As a result of these oxidation reactions, one molecule of GDP is phosphorylated to GTP, one molecule of FAD is reduced to $FADH_2$, and four molecules of NAD^+ are reduced to NADH. Of the four molecules of NADH produced, three come from the citric acid cycle and one from the reaction of the pyruvate dehydrogenase complex. The overall stoichiometry of the oxidation reactions is the sum of the pyruvate dehydrogenase reaction and the citric acid cycle. Note that there is no production of ATP from ADP directly from the citric acid cycle, but many more ATP will arise from reoxidation of NADH and $FADH_2$.

Pyruvate dehydrogenase complex:

$$\text{Pyruvate} + \text{CoA-SH} + NAD^+ \rightarrow \text{acetyl-CoA} + \text{NADH} + CO_2 + H^+$$

Citric acid cycle:

$$\text{Acetyl-CoA} + 3NAD^+ + \text{FAD} + \text{GDP} + \text{Pi} + 2H_2O \rightarrow 2\ CO_2 + \text{CoA-SH} + 3\text{NADH} + 3H^+ + FADH_2 + \text{GTP}$$

Overall reaction:

$$\text{Pyruvate} + 4\ NAD^+ + \text{FAD} + \text{GDP} + \text{Pi} + 2\ H_2O \rightarrow 3CO_2 + 4\ \text{NADH} + FADH_2 + \text{GTP} + 4\ \text{R+}$$

Eventual ATP production per pyruvate:

$$4\ \text{NADH} \rightarrow 10\ \text{ATP}$$

$$1\ FADH_2 \rightarrow 1.5\ ATP$$

$$1GTP \rightarrow 1\ ATP$$

(2.5 ATP for each NADH)

(1.5 ATP for each FADH2)

Total: 12.5 ATP per pyruvate or 25 ATP per glucose

There were also 2 ATP produced per glucose in glycolysis and 2 NADH, which will give rise to another 5 ATP (7 more ATP).

Energetics and control of citric acid cycle

The citric acid cycle itself is also exergonic ($\Delta G°' = -44.3$ kJ $mol^{-1} = -10.6$ kcal mol^{-1}). Of the individual reactions of the cycle, only one is strongly endergonic, the oxidation of malate to oxaloacetate ($\Delta G°' = +29.2$ kJ $mol^{-1} = +7.0$ kcal mol^{-1}). This endergonic reaction is, however, coupled to one of the strongly exergonic reactions of the cycle, the condensation of acetyl-CoA and oxaloacetate to produce citrate and coenzyme A ($\Delta G°' = -32.2$ kj $mol^{-1} = -7.7$ kcal mol^{-1}.

In addition to the energy released by the oxidation reactions, there is more release of energy to come. In the electron transport chain when the four NADH and single $FADH_2$ produced by the pyruvate dehydrogenase complex and citric acid cycle are reoxidized by the electron transport chain, considerable quantities of ATP are produced.

Control of the citric acid cycle is exercised at three points, that is, there are three enzymes within the citric acid cycle that play a regulatory role. There is also a control point outside the cycle itself, namely the reaction in which pyruvate is oxidatively decarboxylated to produce the acetyl-CoA needed for the first reaction of the citric acid cycle proper.

The pyruvate dehydrogenase complex is inhibited by ATP and NADH. Within the citric acid cycle itself the three control points are the reactions catalyzed by citrate synthase, isocitrate dehydrogenase, and the α-ketoglutarate dehydrogenase complex. The first reaction of the cycle is one in which regulatory control appears, as is to be expected in the first reaction of any pathway. Citrate synthase is an allosteric enzyme inhibited by ATP, NADH, and succinyl-CoA. There is again product inhibition by citrate.

The second regulatory site is the isocitrate dehydrogenase reaction. In this case, ADP is an allosteric activator of -the enzyme. We have called attention to the recurring pattern in which ATP and NADH inhibit enzymes of the pathway, and ADP and NAD^+ activate these enzymes. The a-ketoglutarate dehydrogenase complex is the third regulatory site. As before, ATP and NADH are inhibitors and ADP and NAD' are activators.

Succinyl-CoA is also an inhibitor of this reaction. This recurring theme in metabolism reflects the way in which a cell can adjust to an active state or to a resting state. When a cell is metabolically active it uses ATP and NADH at a great rate, producing large amounts of ADP and NAD^+. In other words, when the ratio ATP/ADP is low, the cell is using energy and needs to release more energy from stored nutrients.

A low $NADH/NAD^+$ ratio is also characteristic of an active metabolic state. On the other hand, a resting cell has fairly high levels of ATP and NADH. The ratios of ATP/ADP and $NADH/NAD^+$ are also high in resting cells, which do not need to maintain a high level of oxidation to produce energy. When cells have low energy requirements (they have a high "energy charge") with high ATP/ADP and $NADH/NAD^+$ ratios, the presence of so much ATP and NADH serves as a signal to "shut down" the enzymes responsible for oxidative reactions. When cells have a low energy charge, characterized by low ATP/ADP and NADH/NAD^+ ratios, the need to release more energy and to generate more ATP serves as a signal to "turn on" the oxidative enzymes.

The glyoxilate cylce

In plants and in some bacteria, but not in animals, acetyl-CoA can serve as the starting material for the biosynthesis of carbohydrates. Animals can convert carbohydrates to fats, but not fats to carbohydrates. Two enzymes are responsible for the ability of plants and bacteria to produce glucose from fatty acids. Isocitratelyase cleaves isocitrate, producing glyoxylate and succinate. Malate synthase catalyzes the reaction of glyoxylate with acetyl-CoA to produce malate. These two reactions in succession bypass the two oxidative decarboxylation steps of the citric acid cycle. The net result is an alternative pathway, the glyoxylate cycle.

Two molecules of acetyl-CoA enter the glyoxylate cycle; they give rise to one molecule of malate and eventually to one molecule of oxaloacetate. Two 2-carbon units (the acetyl groups of acetyl-CoA) give rise to a 4-carbon unit (malate), which is then converted to oxaloacetate (also a 4-carbon compound). Glucose can then be produced from oxaloacetate by gluconeogenesis. Specialized organelles in plants, called glyoxysomes, are the sites of the glyoxylate cycle. This pathway is particularly important in germinating seeds. The fatty acids stored in the seeds are broken down for energy during germination. First the fatty acids give rise to acetyl-CoA, which can enter the citric acid cycle and go on to release energy in the ways we have already seen.

This pathway results in the net conversion of two acetyl-CoA to oxaloacetate. The unique reactions of the glyoxylate cycle are shown in green, and the main features of the citric acid cycle are shown in blue. All metabolic pathways are related and all operate simultaneously. In the case of catabolic pathways, nutrients, many of which are macromolecules, are broken down to smaller molecules such as sugars, fatty acids, and amino acids. These small molecules are processed further, and the end products of catabolism frequently enter the citric acid cycle, which plays a key role in metabolism.

The citric acid cycle and the glyoxylate cycle can operate simultaneously. Acetyl-CoA also serves as the starting point for the synthesis of glucose and any other compounds needed by the growing seedling. The glyoxylate cycle also occurs in bacteria. This point is far from surprising, since many types of bacteria can live on very limited carbon sources. They have metabolic pathways that can produce all the biomolecules they need from quite simple molecules. The glyoxylate cycle is one example of how bacteria manage this feat.

Citric acid cycle in catabolism

The nutrients taken in by an organism can include large molecules. This observation is especially true in the case of animals, which ingest polysaccharides and proteins, which are polymers, as well as lipids. Nucleic acids constitute a very small percentage of the nutrients present in foodstuffs, and we shall not consider their catabolism.

The first step in the breakdown of nutrients is the degradation of large molecules to smaller ones. Polysaccharides are hydrolyzed by specific enzymes to produce sugar monomers; an example is the breakdown of starch by amylases. Lipases hydrolyze triacylglycerols to give fatty acids and glycerol. Proteins are digested by proteases, with amino acids as the end products. Sugars, fatty acids, and amino acids then enter their specific catabolic pathways.The catabolic reactions occur in the cytosol; the citric acid cycle takes place in mitochondria. Many of the end products of catabolism cross the mitochondrial membrane and then participate in the citric acid cycle. This figure also shows the outline of pathways by which amino acids are converted to components of the citric acid cycle.

Citric acid cycle in anabolism

The citric acid cycle is a source of starting materials for the biosynthesis of many important biomolecules, but the supply of the starting materials that are components of the cycle must be replenished if the cycle is to continue operating. In particular, the oxaloacetate in an organism must be maintained at a level sufficient to allow acetyl-CoA to enter the cycle. In some organisms acetyl-CoA can be converted to oxaloacetate and other citric acid cycle intermediates by the glyoxylate cycle, but mammals can not do this.

In mammals oxaloacetate is produced from pyruvate by the enzyme pyruvate carboxylase. We already encountered this enzyme and this reaction in the context of gluconeogenesis and here we have another highly important role for this enzyme and the reaction it catalyzes. The supply of oxaloacetate would soon be depleted if there were no means of producing it from a readily available precursor. This reaction, which produces oxaloacetate from pyruvate, provides a connection between the amphibolic citric acid cycle and the anabolism of sugars by gluconeogenesis.

On this same topic of carbohydrate anabolism, we should note again that pyruvate cannot be produced from acetyl-CoA in mammals. Since acetyl-CoA is the end product of catabolism of fatty acids, we can see that mammals could not exist with fats or acetate as the sole carbon source. The intermediates of carbohydrate metabolism would soon be depleted. Carbohydrates are the principal energy and carbon source in animals. Plants can carry out the conversion of acetyl-CoA to pyruvate and oxaloacetate, so they can exist without carbohydrates as a carbon source.

The conversion of pyruvate to acetyl-CoA does take place in both plants and animals. The anabolic reactions of gluconeogenesis take place in the cytosol. Oxaloacetate is not transported across the mitochondrial membrane. Two mechanisms exist for the transfer of molecules needed for gluconeogenesis from mitochondria to the cytosol. One mechanism takes advantage of the fact that phosphoenolpyruvate can be formed from oxaloacetate in the mitochondrial matrix; phosphoenolpyruvate is then transferred to the cytosol, where the remaining reactions take place. The other mechanism relies on the fact.

The citric acid cycle is important not only as a source of energy during aerobic metabolism but also as a key pathway in the synthesis of important metabolic intermediates. We shall see in subsequent chapters that it is a source of starting materials for the production of amino acids, carbohydrates, vitamins, nucleotides, and heme. However, if these intermediates are used for the synthesis of other molecules, then they must be replenished to maintain the catalytic nature of this cycle.

The term anaplerotic means "filling up," and the reactions that replenish the citric acid cycle are called anaplerotic reactions. One source of needed compounds, available to all organisms, is that group of amino acids that can be converted to citric acid cycle intermediates in a single reaction. A simple reaction available to all organisms is to add carbon dioxide to the pyruvate and phosphoenolpyruvate generated from metabolism of sugars.

Some anaerobic organisms have developed only parts of the citric acid cycle, which they use exclusively to make the important precursors. These simple yet important reactions emphasize the truly connected nature of what we often artificially separate into "pathways." They also illustrate the convergence of evolution to a few key molecules and metabolic steps that malate, which is another intermediate of the citric acid cycle, can be transferred to the cytosol. There is a malate dehydrogenase enzyme in the cytosol as well as in mitochondria, and malate can be converted to oxaloacetate in the cytosol.

$$\text{Malate} + \text{NAD}^+ \rightarrow \text{oxaloacetate} + \text{NADH} + \text{H}^+$$

Oxaloacetate is then converted to phosphoenolpyruvate, leading to the rest of the steps of gluconeogenesis. Gluconeogenesis has many steps in common with the production of glucose in photosynthesis, but photosynthesis also has many reactions in common with the pentose phosphate pathway. Thus, nature has evolved common strategies to deal with carbohydrate metabolism in all its aspects.

Lipid Anabolism

The anabolic reactions of lipid metabolism, like those of carbohydrate metabolism, take place in the cytosol; these reactions are catalyzed by soluble enzymes that are not bound to membranes. Acetyl-CoA is mainly produced in mitochondria, whether from pyruvate or from breakdown of fatty acids. An indirect transfer mechanism exists for transfer of acetyl-CoA in which citrate is transferred to the cytosol.

Citrate reacts with CoA-SH to produce citryl-CoA, which is then cleaved to yield oxaloacetate and acetyl-CoA. The enzyme that catalyzes this reaction requires ATP and is called ATP-citratelyase. The overall reaction is

$$\text{Citrate} + \text{CoA-SH} + \text{ATP} \rightarrow \text{acetyl-CoA} + \text{oxaloacetate} + \text{ADP} + \text{Pi}$$

Acetyl-CoA is the starting point for lipid anabolism in both plants and animals. Oxaloacetate can be reduced to malate by the reverse of a reaction we saw in the last section in the context of carbohydrate anabolism.

$$\text{Oxaloacetate} + \text{NADH} + H^+ \rightarrow \text{malate} + \text{NAD}^+$$

Malate can move into and out of mitochondria by active transport processes, and the malate produced in this reaction can be used again in the citric acid cycle. However, malate need not be transported back into mitochondria but can be oxidatively decarboxylated to pyruvate by malic enzyme, which requires NADP+.

$$\text{Malate} + \text{NADP}^+ \text{®} \text{ pyruvate} + CO_2 + \text{NADPH} + H^+$$

Tranjer of the starting materials of lipid anabolism from the mitochrondrion to the cytosol. It is not definitely established whether acetyl-CoA is transported from the mitochondrion to the cytosol.

Reactions involving citric acid cycle intermediates that produce NADPH for fatty acid anabolism. Note that these reactions take place in the cytosol. These reactions are a reduction reaction followed by an oxidation; there is no net oxidation. There is, however, a substitution of NADPH for NADH. This last point is an important one, since many of the enzymes of fatty acid synthesis require NADPH. The pentose phosphate pathway is the principal source of NADPH in most organisms, but here we have another source as well. The two ways of producing NADPH clearly indicate that all metabolic pathways are related. The exchange reactions involving malate and citryl-CoA constitute a control mechanism in lipid anabolism, whereas the pentose phosphate pathway is part of carbohydrate metabolism. Both carbohydrates and lipids are important energy sources in many organisms, particularly animals.

Anabolic reactions that produce amino acids have, as a starting point, those intermediates of the citric acid cycle that can cross the mitochondrial membrane into the cytosol. We have already seen that malate can cross the mitochondrial membrane and give rise to oxaloacetate in the cytosol. Oxaloacetate can undergo a transamination reaction to produce aspartate, and aspartate in turn can undergo further reactions to form not only amino acids but also other nitrogen-containing metabolites, such as pyrimidines. Similarly, isocitrate can cross the mitochondrial membrane and produce α-ketoglutarate in the cytosol.

Glutamate arises from α-ketoglutarate as a result of another transamination reaction, and glutamate undergoes further reactions to form still more amino acids. Succinyl-CoA gives rise not to amino acids but to the porphyrin ring of the heme group. Another difference is that the first reaction of heme biosynthesis, the condensation of succinyl-CoA and glycine to form δ-aminolevulinic acid, takes place in the mitochondrial matrix, whereas the remainder of the pathway occurs in the cytosol.

The operation of any metabolic pathway, anabolic or catabolic, can be "speeded up" or "slowed down" in response to the needs of an organism by control mechanisms such as feedback control. Regulation of metabolism takes place in similar ways in many different pathways. The citric acid cycle is considered part of aerobic metabolism, but we have not encountered any reactions in this chapter in which oxygen takes part. The reactions of the citric acid cycle are intimately related to those of electron transport and oxidative phosphorylation, which do eventually lead to oxygen. The citric acid cycle provides a vital link between the chemical energy of nutrients and the chemical energy of ATP.

Chapter 6

Nitrogen Metabolism

In the metabolic pathways of nitrogen, the first thing to discuss is the nitrogen fixation, the process by which inorganic molecular nitrogen from the atmosphere NO is incorporated first into ammonia and then into organic compounds that are of use to organisms. Nitrate ion (NO_3^-), another kind of inorganic nitrogen, is the form in which nitrogen is found in the soil, and many fertilizers contain nitrates, frequently potassium nitrate.

The process of nitrification (nitrate reduction to ammonia) provides another way for organisms to obtain nitrogen. Nitrate ion and nitrite ion (NO_3^-), are also involved in denitrification reactions, which return nitrogen to the atmosphere. Ammonia formed by either pathway, nitrogen fixation or nitrification, enters the biosphere. Ammonia is converted to organic nitrogen by plants, and organic nitrogen is passed to animals through food chains. Finally, waste products such as urea are excreted and degraded to ammonia by microorganisms.

The word "ammonia" comes from sal ammoniac: (ammonium chloride), which was first prepared from the dung of camels at the temple of Jupiter Ammon in North Africa. The process of death and decay releases ammonia in both plants and animals. Denitrifying bacteria reverse the conversion of ammonia to nitrate and then recycle the NO_3^- as free N_2.

The topic of nitrogen metabolism includes the biosynthesis and breakdown of amino acids, purines, and pyrimidines; also, the metabolism of porphyrins is related to that of amino acids. Many of these pathways, particularly anabolic ones, are long and complex. Where the amount of material is large and highly detailed, we shall concentrate on the most important points. Specifically, we shall concentrate on overall patterns and on interesting reactions of wide applicability.

Nitrogen fixation

Bacteria are responsible for the reduction of N_2 to ammonia (NH_3). Typical nitrogen-fixing bacteria are symbiotic organisms that form nodules on the roots of leguminous plants such as beans and alfalfa. Many free-living microbes and some cyanobacteria also fix nitrogen. Plants and animals cannot carry out nitrogen fixation. This conversion of molecular nitrogen to ammonia is the only source of nitrogen in the biosphere except for that provided by nitrates. The conjugate acid form of NH_3, ammonium ion (NH_4), is the form of nitrogen that is used in the first stages of the synthesis of organic compounds.

Parenthetically, NH_3 obtained by chemical synthesis from nitrogen and hydrogen is the starting point for the production of many synthetic fertilizers, which frequently contain nitrates. The nitrogenase enzyme complex found in nitrogen-fixing bacteria catalyzes the production of ammonia from molecular nitrogen.

The half reaction of reduction is

$$N_2 + 16e^- + 12ATP + 12\ H_2O + 8H+ \rightarrow 1NH_4^+ + 12ADP + 12\ Pi$$

The half reaction of oxidation will vary, because different

organisms vary in terms of the substance oxidized to supply electrons. Several proteins are included in the nitrogenase complex. Ferredoxin is one of them. There are also two proteins specific to the nitrogenase reaction.

One is an iron-molybdenum (Fe-Mo) protein, and the other is an iron-sulphur protein, usually referred to as the Fe protein. The flow of electrons is from ferredoxin to the Fe protein to the Fe-Mo protein to nitrogen.

The nature of the nitrogenase complex is a subject of active research. Significant progress has been made in this work with the determination by x-ray crystallography of the three-dimensional structure of both the Fe protein and the Fe-Mo protein from Azotobacter vinelandii. The Fe protein is a dimer ("the iron butterfly"), with the iron-sulphur cluster located at the butterfly's head.

Crop production in the United States is higher than in many areas of the world. In part, this is because of the extensive use of fertilizers, especially those that supply nitrogen in a form that plants can use readily. Both ammonium and nitrate ions are used; even ammonia gas can be pumped into the ground if there is enough water available to dissolve it. Since ammonia is toxic to animals, it often comes as a surprise that ammonia gas itself can be used for fertilization. Plants can assimilate it rapidly, but usually they never get the chance because the nitrifying soil bacteria, especially Nitrosomonas and Nitrobacter, rapidly convert the ammonia to nitrite and then to nitrate, respectively. The final nitrate product is easily converted back to ammonia, but it takes energy to do it.

Ammonia is especially useful as a fertilizer in the early spring and for germinating plants. In the spring, the soil is usually damp enough to dissolve the ammonia so it can move to the plants. Because light is less available in the early spring, the young plants do not have enough energy to convert the nitrate back to ammonia until their chloroplasts develop fully.

Fortunately, the ammonia has gone directly to the plants rather than to the soil bacteria because of the condition of the soil. The genes for the enzymes for nitrogen fixation have been studied extensively. There is much research going on to determine if these genes could be incorporated into crop plants, which would reduce the amount of nitrogen fertilizer needed for maximal plant growth and crop production.

Feedback control

The biosynthetic pathways that produce amino acids and nucleobases are long and complex, requiring a large investment of energy by the organism. If there is a high level of some end product, such as an amino acid or a nucleotide, the cell saves energy by not making that compound. However, the cell needs some signal not to produce more of that particular compound. The signal is frequently a feedback inhibition,mechanism, in which the end product of a metabolic pathway inhibits an enzyme at the beginning of the pathway.

Alllosteric enzyme aspartate transcarbamoylase is the enzyme catalyzes one of the early stages of pyrimidine nucleotide biosynthesis, and it is inhibited by the end product of that pathway, namely cytidine triphosphate (CTP). Feedback inhibition is frequently encountered in the biosynthesis of amino acids and nucleotides.

Metabolism of aminoacids

Ammonia is toxic in high concentrations, and so it must be incorporated into biologically useful compounds when it is formed by the reactions of nitrogen fixation. The amino acids glutamate and glutamine are of central importance in the process. Glutamate arises from α-ketoglutarate, and glutamine from glutamate.

The production of glutamate is a reductive amination, and the production of glutamine is amination. In other reactions of amino acid anabolism the α-amino group of glutamate and the side-chain amino group of glutamine are shifted to other compounds in transamination reactions.

Biosynthesis of amino acids involves a common set of reactions. In addition to transamination reactions, transfer of one-carbon units, such as formyl or methyl groups, occurs frequently. We are not going to discuss all the details of the reactions that give rise to amino acids. We can, however, organize this material by grouping amino acids into families based on common precursors. The reactions of some of the individual families of amino acids provide good examples of those reactions that are of general importance, transamination and one-carbon transfer.

We can also make some generalizations about amino acid metabolism in terms of the relationship of the carbon skeleton to the citric acid cycle and the related reactions of pyruvate and acetyl-CoA. The citric acid cycle is amphibolic; it has a part in both catabolism and anabolism. The anabolic aspect of the citric acid cycle is of interest in amino acid biosynthesis.

The catabolic aspect is apparent in the breakdown of amino acids, leading to their eventual excretion, which takes place in reactions related to the citric acid cycle. Glutamate is formed from NH_4^+ and α-ketoglutarate in a reductive amination and that requires NADPH. This reaction is reversible and is catalyzed by glutamate dehydrogenase (GDH).

$$NH_4^+ \ \alpha\text{-ketoglutarate} + NADPH + H^+ \rightarrow \text{glutamate} + NADP^+ + H_2O$$

Glutamate is a major donor of amino groups in reactions, and α-ketoglutarate is a major acceptor of amino groups. Note the requirement for reducing power. The conversion of glutamate to glutamine is catalyzed by glutamine synthetase (GS) in a reaction that requires ATP.

$$NH^+_4 + \text{glutamate} + ATP \rightarrow \text{glutamine} + ADP + Pi$$

These reactions fix inorganic nitrogen (NH_3), forming organic (carbon-containing) nitrogen compounds, such as amino acids, but they frequently do not operate in this sequential fashion. When nitrogen is limiting, as is frequently the case in plants (fertilizers list the amount of nitrogen they contain), the conversion of glutamate to glutamine is the preferred mode of nitrogen assimilation. This means that the supply of glutamate will become depleted unless there is some way of replenishing it. The reductive amination of α-ketoglutarate with the amide nitrogen of glutamine as the nitrogen source is the way this is done.

$$\text{Reluctant} + \alpha\text{-ketoglutarate} + \text{glutamine} \rightarrow 2 \text{ glutamate} + \text{oxidized reductant}$$

The reductant can be NADH or NADPH (in yeast and bacteria) or reduced ferredoxin (in plants). The enzyme that catalyzes this reaction is glutamine synthase; it is also known as glutamate:oxo-glutarate amino-transferase, a name that gives rise to the acronym GOGAT. A GS/GOGAT complex exists in plants and allows them to cope with conditions of limited nitrogen availability. Quite frequently, enzymes that catalyze transamination reactions require pyridoxal phosphate as a coenzyme.

The role of pyridoxal phosphate in transamination reactions. (a) The mode of binding of pyridoxal phosphate (PyrP) to the enzyme (E) and to the substrate amino acid. (b) The reaction itself. The original substrate, an amino acid, is deaminated, while an aceto acid is aminated to form an amino acid. The net reaction is one of transamination. Note that the coenzyme is regenerated and that the original substrate and final product are both amino acids. Pyridoxal phosphate forms a Schiff base with the amino group of Substrate I (the amino group donor).

The next stage is a rearrangement followed by hydrolysis, which removes Product I (the α-keto acid corresponding to Substrate 1). The coenzyme now carries the amino group (pyridoxamine). Substrate II (another a-keto acid) then forms a Schiff base with pyridoxamine.

Again there is a rearrangement followed by a hydrolysis, which gives rise to Product II (an amino acid) and regenerates pyridoxal phosphate. The net reaction is that an amino acid (Substrate 1) reacts with an α-keto acid (Substrate II) to form an a-keto acid (Product 1) and an amino acid (Product II). The amino group has been transferred from Substrate I to Substrate II, forming the amino acid, Product II.

In addition to transamination reactions, one-carbon transfer reactions occur frequently in amino acid biosynthesis. A good example of a one-carbon transfer can be found in the reactions that produce the amino acids of the serine family. This family also includes glycine and cysteine.

Serine and glycine themselves are frequently precursors in other biosynthetic pathways. The ultimate precursor of serine is 3-phosphoglycerate, which is obtainable from the glycolytic pathway. The hydroxyl group on carbon 2 is oxidized to a keto group, giving an α-keto acid.

A transamination reaction in which glutamate is the nitrogen donor produces 3-phosphoserine. This reaction is an example of a transamination involving glutamate and a-ketoglutarate. Hydrolysis of the phosphate group then gives rise to serine. The conversion of serine to glycine involves the transfer of a one-carbon unit from serine to an acceptor. This reaction is catalyzed by seyine hydroxymethylase, with pyridoxal phosphate as a coenzyme.

The acceptor in this reaction is tetrahydrofolate, a derivative of folic acid and a frequently encountered carrier of one-carbon units in metabolic pathways. Its structure has three parts: a substituted pteridine ring, p-aminobenzoic acid, and glutamic acid. Folic acid is a vitamin that has been identified as essential in preventing birth defects; consequently, it is now recommended for all women of childbearing age. There is also some evidence that folic acid may prevent heart disease in both men and women older than 50.

$$\text{Serine} + \text{tetrahydrofolate} \rightarrow \text{glycine} + \text{methylenetetrahydrofolate} + H_2O$$

The one-carbon unit transferred in this reaction is bound to tetrahydrofolate, forming N^5 N^{10}-methylenetetra-hydrofolate, in which the methylene (one-carbon) unit is bound to two of the nitrogens of the carrier. Tetrahydrofolate is not the only carrier of one-carbon units. We have already encountered biotin, a carrier of CO_2 and we have discussed the role that biotin plays in gluconeogenesis and in the anabolism of fatty acids.

The conversion of serine to cysteine involves some interesting reactions. The source of the sulphur in animals differs from that in plants and bacteria. In plants and bacteria, serine is acetylated to form a-acetylserine. This reaction is catalyzed by serine acyltransferase, with acetyl-CoA as the acyl donor.

Conversion of a-acetylserine to cysteine requires production of sulphide by a sulphur donor. The sulphur donor for plants and bacteria is 3'-phospho-5'-adenyl sulphate.

The sulphate group is reduced first to sulphite and then to sulphide. The sulphide, in the conjugate acid form HS^- displaces the acetyl group of the a-acetylserine to produce cysteine. Animals form cysteine from serine by a different pathway, since they do not have the enzymes to carry out the sulphate to sulphide conversion that we have just seen. The reaction sequence in animals involves the amino acid methionine.

Methionine, which is produced by reactions of the aspartate family in bacteria and plants, cannot be produced by animals. It must be obtained from dietary sources. It is an essential amino acid because it cannot be synthesized by the body.

The ingested methionine reacts with ATP to form S-adenosylrnethionine, which has a highly reactive methyl group. This compound is a carrier of methyl groups in many reactions. The methyl group from α-adenosylmethionine can be transferred to any one of a number of acceptors, producing sadenosyl-homocysteine.

Hydrolysis of α-adenosylhomocysteine in turn produces homocysteine. Cysteine can be synthesized from serine and homocysteine, and this pathway for cysteine biosynthesis is the only one available to animals. Serine and homocysteine react to produce cystathionine, which hydrolyzes to form cysteine, NH_4, and α-ketobutyrate. It is worth noting that we have now seen three important carriers of one-carbon units: biotin, a carrier of CO_2; tetrahydrofolate (FH_4), a carrier of methylene and formyl groups; and S-adenosylmethionine, a carrier of methyl groups.

Essential aminoacids

Biosynthesis of proteins requires the presence of all the constituent amino acids. If 1 of the 20 amino acids is missing or in short supply, protein biosynthesis is inhibited. Some organisms, such as E. coli, can synthesize all the amino acids that they need. Other species, including humans, must obtain some amino acids from dietary sources. The body can synthesize some of these amino acids, but not in sufficient quantities for its needs, especially in the case of growing children. This last point applies particularly to children's requirement for arginine and histidine.

Amino acids are not stored (except in proteins), and dietary sources of essential amino acids are needed at regular intervals. Protein deficiency, especially a prolonged deficiency in sources that contain essential amino acids, leads to the disease kwashiorkor. The problem in this disease, particularly severe in growing children, is not simply starvation, but the breakdown of the body's own proteins.

The urine cycle

A central pathway in nitrogen metabolism is the urea cycle. It plays a role in both catabolism and anabolism. The nitrogens that enter the urea cycle do so first as ammonia in the form of ammonium ion. The immediate precursor is glutamate, but the ammonia nitrogens of glutamate have ultimately come from many sources as a result of transamination reactions.

A condensation reaction between the ammonium ion and carbon dioxide produces carbamoyl phosphate in a reaction that requires two molecules of ATP for each molecule of carbamoyl phosphate. Carbamoyl phosphate reacts with ornithine to form citrulline. The reactions of the cycle up to this point take place in the mitochondrion. Citrulline is then transported to the cytosol. A second nitrogen enters the urea cycle when aspartate reacts with citrulline to form argininosuccinate in another reaction that requires ATP.

The amino group of the aspartate is the source of the second nitrogen in the urea that will be formed in this series of reactions. Argininosuccinate is split to produce arginine and fumarate. Finally, arginine is hydrolyzed to give urea and to regenerate ornithine, which is transported back to the mitochondrion. Another way of looking at the urea cycle is to consider arginine the immediate precursor of urea and to see it as producing ornithine in the process. According to this point of view, the rest of the cycle is the regeneration of arginine from ornithine.

The synthesis of famarate is a link between the urea cycle and the citric acid cycle. Fumarate is, of course, an intermediate of the citric acid cycle, and it can be converted to oxaloacetate. A transamination reaction can convert oxaloacetate to aspartate, providing another link between the two cycles. Four "high-energy" phosphate bonds are required because of the production of pyrophosphate in the conversion of aspartate to argininosuccinate. When we specifically focus on the catabolism of amino acids, the first step we consider is the removal of nitrogen by transamination.

Transamination reactions are also important in the anabolism of amino acids, so it is important to remind ourselves that anabolic and catabolic pathways are not the exact reverse of each other, nor do they involve exactly the same group of enzymes. In catabolism, the amino nitrogen of the original amino acid is transferred to α-ketoglutarate to produce glutamate, leaving behind the carbon skeleton. The fates of the carbon skeleton and of the nitrogen can be considered separately. Breakdown of the carbon skeleton of amino acids follows two general pathways, the difference between the two pathways depending on the type of end product.

A glucogenic amino acid is one that yields pyruvate or oxaloacetate on degradation. Oxaloacetate is the starting point for the production of glucose by gluconeogenesis. A ketogenic amino acid is one that breaks down to acetyl-CoA or acetoacetyl-CoA, leading to the formation of ketone bodies. The carbon skeletons of the amino acids give rise to metabolic intermediates such as pyruvate, acetyl-CoA, acetoacetyl-CoA, cc-ketoglutarate, succinyl-CoA, fumarate, and oxaloacetate.

Oxaloacetate is a key intermediate in the breakdown of the carbon skeleton of amino acids because of its dual role in the citric acid cycle and in gluconeogenesis. The amino acids degraded to acetyl-CoA and acetoacetyl-CoA are used in the citric acid cycle, but mammals cannot synthesize glucose from acetyl-CoA. This fact is the source of the distinction between glucogenic and ketogenic amino acids.

Glucogenic amino acids can be converted to glucose, with oxaloacetate as an intermediate, but ketogenic amino acids cannot. The nitrogen portion of amino acids is involved in transamination reactions in breakdown as well as in biosynthesis. Excess nitrogen is excreted in one of three forms: ammonia (as ammonium ion), urea, and uric acid.

Animals, such as fish, that live in an aquatic environment excrete nitrogen as ammonia; they are protected from the toxic effects of high concentrations of ammonia not only by the removal of ammonia from their bodies but also by rapid dilution of the excreted ammonia by the water in the environment.

The principal waste product of nitrogen metabolism in terrestrial animals is urea (a water-soluble compound); its reactions provide some interesting comparisons with the citric acid cycle. Birds excrete nitrogen in the form of uric acid, which is insoluble in water. They do not have to carry the excess weight of water, which could hamper flight, to rid themselves of waste products.

Ammonia gas is toxic to most organisms and must usually be disposed of rapidly. In a certain sense, you can almost guess the mechanism of nitrogen waste disposal if you know the amount of water available to the organism. For example, bacteria and fish, which live in "infinite" water supplies, usually simply release ammonia into the medium, where it can be used by organisms that are lower on the evolutionary scale.

Fish sometimes produce trimethylamine, another highly water-soluble compound, which is the characteristic "fish odour." Most terrestrial animals do not have "infinite" water supplies, but the mammals have a bladder and usually live in conditions where adequate water is available. Their mechanism for disposal of most toxins is to prepare a water-soluble compound and then excrete it through the urine.

Thus, urea becomes a major by product of nitrogen metabolism in mammals. Reptiles and other desert animals do not usually have much water available, and birds cannot afford the luxury of the weight of a bladder.

These animals do not make urea; rather, they

convert all their waste nitrogen to uric acid, the concentrated white solid so familiar in bird droppings. Some desert mammals, such as the kangaroo rat, which never drinks water but rather lives off metabolic water, also convert some of their waste nitrogen to uric acid to conserve the water used in urine.

Uric acid, the typical waste product from purines, can cause problems in primates because of its marginal water solubility. Deposits of uric acid in the joints and extremities cause gout. Other mammals do not have a problem with uric acid, since they convert it to allantoin and then to allantoic acid that is either excreted (in some fish) or further catabolized to urea or ammonia.

Anobolism of purine nucleotides

The biosynthetic pathway for both purine and pyrimidine nucleotides makes use of preformed ribose-5-phosphate. Purines and pyrimidines are synthesized in different ways, and we shall consider them separately.

Anabolism of Inosine Monophosphate (IMP)

In the synthesis of purine nucleotides, the growing ring system is bonded to the ribose phosphate while the purine skeleton is being assembled, first the five-membered ring and then the six-membered ring, eventually producing inosine 5'-monophosphate.

All four nitrogen atoms of the purine ring are derived from amino acids: two from glutamine, one from aspartate, and one from glycine. Two of the five carbon atoms (adjacent to the glycine nitrogen) also come from glycine, two more come from tetrahydrofolate derivatives, and the fifth comes from CO_2 IMP is the precursor of both AMP and GMP. The conversion of IMP to AMP takes place in two stages. The first step is the reaction of aspartate with IMP to form adenylosuccinate.

This reaction is catalyzed by adenylosuccinate synthetase and requires GTP, not ATP.

The cleavage of fumarate from adenylosuccinate to produce AMP is catalyzed by adenylosuccinase. This enzyme also functions in the synthesis of the six-membered ring of IMP. The conversion of IMP to GMP also takes place in two stages. The first of the two steps is an oxidation in which the C-H group at the C-2 position is converted to a keto group. The oxidizing agent in the reaction is $NAD^{+,}$ and the enzyme involved is IMP dehydrogenase. The nucleotide formed by the oxidation reaction is xanthosine 5'-phosphate (XMP).

An amino group from the side chain of glutamine replaces the C-2 keto group of XMP to produce GMP. This reaction is catalyzed by GMP synthetase; ATP is hydrolyzed to AMP and PPj in the process. Note that there is some control over the relative levels of purine nucleotides; GTP is needed for the synthesis of adenine nucleotides, whereas ATP is required for the synthesis of guanine nucleotides. Each of the purine nucleotides must occur at a reasonably high level for the other to be synthesized. Subsequent phosphorylation reactions produce purine nucleoside diphosphates (ADP and GDP) and triphosphates (ATP and GTP).

The purine nucleoside monophosphates, diphosphates, and triphosphates are all feedback inhibitors of the first stages of their own biosynthesis. Also, AMP, ADP, and ATP inhibit the conversion of IMP to adenine nucleotides, and GMP, GDP, and GTP inhibitable conversion of IMP to xanthylate and to guanine nucleotides.

The production of IMP starting with ribose-5-phosphate requires the equivalent of seven ATP. The conversion of IMP to AMP requires hydrolysis of an additional "high-energy" bond in this case, that of GTP.

In the formation of AMP from ribose-5-phosphate, the equivalent of eight ATP is needed. The conversion of IMP to GMP requires two "high-energy" bonds, since a reaction occurs in which ATP is hydrolyzed to AMP and PPj. For the production of GMP from ribose-5-phosphate, the equivalent of nine ATP is necessary.

The anaerobic oxidation of glucose produces only two ATP for each molecule of glucose. Anaerobic organisms require four molecules of glucose (which produce eight ATP) for each AMP they form, or five molecules of glucose (which produce ten ATP) for each GMP. The process is more efficient for aerobic organisms. Since 30 or 32 ATP result from each molecule of glucose, depending on the type of tissue, aerobic organisms can optimally produce four AMP (requiring 32 ATP) or three GMP (requiring 27 ATP) for each molecule of glucose oxidized.

A mechanism for reuse of purines, rather than complete turnover and new synthesis, saves energy for organisms. The biosynthesis of nucleotides requires considerable expenditures of energy by organisms in long and complex pathways, and, as a result, salvage reactions exist for reuse of some of these compounds.

Catabolic reactions of purine nucleiotides

The catabolism of purine nucleotides proceeds by hydrolysis to the nucleoside and subsequently to the free base, which is further degraded. Deamination of guanine produces xanthine, and deamination of adenine produces hypoxanthine. Hypoxanthine can be oxidized to xanthine, so this base is a common degradation product of both adenine and guanine. Xanthine is oxidized in turn to uric acid.

In birds, some reptiles, and insects, and in Dalmatian dogs and primates (including humans), uric acid is the end product of purine metabolism and is excreted. In all other terrestrial animals, including all other mammals, allantoin is the product excreted, whereas allantoate is the product in fish. Allantoate is further degraded to glyoxylate and urea by microorganisms and some amphibia.

Gout is a disease in humans that is caused by overproduction of uric acid. Deposits of uric acid (which is barely soluble in water) accumulate in the joints of the hands and feet. Benjamin Franklin suffered from gout, and he left eloquent testimony to the painful nature of this disease.

Allopurinol is a compound used to treat gout; it inhibits the degradation of hypoxanthine to xanthine and of xanthine to uric acid, preventing the buildup of uric acid deposits. Salvage reactions are important in the metabolism of purine nucleotides because of the amount of energy required for the synthesis of the purine bases.

A free purine base that has been cleaved from a nucleotide can produce the corresponding nucleotide by reacting with the compound phosphoribosylpyrophosphate (PRPP), formed by a transfer of a pyrophosphate group from ATP to ribose-5-phosphate.

Two different enzymes with different specificities with respect to the purine base catalyze salvage reactions. The reaction

$$\text{Adenine} + \text{PRPP} \rightarrow \text{AMP} + \text{PPi}$$

is catalyzed by adenine phosphoribosyltransferase. The corresponding reactions of guanine and hypoxanthine

$$\text{Hypoxanthine} + \text{PRPP} \xrightarrow{\text{HPRT}} \text{IMP} + \text{Ppi}$$

$$\text{Guanine} + \text{PRPP} \xrightarrow{\text{HPRT}} \text{GMP} + \text{PPi}$$

are catalyzed by hypoxanthine-guanine phosphoribosyltransferase (HPRT). A deficiency in HPRT can result in a serious disorder, the Lesch-Nyhan syndrome.

A deficiency of the HPRT enzyme is the cause of the Lesch-Nyhan syndrome, a genetic disease. The biochemical consequences include an elevated concentration of PRPP and increased production of purines and uric acid. The accumulation of uric acid leads to kidney stones and gout, but the most striking clinical manifestations are neurological in nature.

There is a compulsive tendency toward self-mutilation among patients with the Lesch-Nvhan syndrome; they tend to bite off fingertips and parts of the lips. The development of kidney stones and gouty symptoms can be prevented by the administration of allopurinol, but there is no real treatment for the self-destructive behaviour and the mental retardation and spasticity that accompany it.

The diverse manifestations of this disease show clearly that metabolism is extremely complex and that the failure of one enzyme to function can have consequences that reach far beyond the reaction that it catalyzes.

Pyrimidine nucleotide metabolism

Overall scheme of pyrimidine nucleotide biosynthesis differs from that of purine nucleotides in that the pyrimidine ring is assembled before it is attached to ribose-5-phosphate. The carbon and nitrogen atoms of the pyrimidine ring come from carbamoyl phosphate and aspartate. The production of carbamoyl phosphate for pyrimidine biosynthesis takes place in the cytosol, and the nitrogen donor is glutamine.

$$HCO_3^- + \text{glutamine} + 2ATP + H_2O \rightarrow \text{carbamoyl phosphate} + \text{glutamate} + 2\ ADP + Pi$$

The reaction of carbamoyl phosphate with aspartate to produce N-carbamoylaspartate is the committed step in pyrimidine biosynthesis. The compounds involved in reactions up to this point in the pathway can play other roles in metabolism; after this point, N-carbamoylaspartate can be used only to produce pyrimidines thus, the term "committed step."

The next step, the conversion of N-carbamoyl-aspartate to dihydroorotate, takes place in a reaction that involves an intramolecular dehydration (loss of water) as well as cyclization. This reaction is catalyzed by dihydroorotase. Dihydroorotate is converted to orotate by dihydroorotate clehydrogenase, with the concomitant conversion of NAD^+ to NADH.

A pyrimidine nucleotide is now formed by the reaction of orotate with PRPP to give oroticline 5'-phosphate (OMP), which is a reaction similar to the one that takes place in purine salvage. Orotate phosphoribosyl transferase catalyzes this reaction. Finally, orotidine 5'-phosphate decarboxylase catalyzes the conversion of OMP to UMP (uridine 5'-phosphate), which is the precursor of the remaining pyrinlicline nucleotides.

Two successive phosphorylation reactions convert UMP to UTP. The conversion of uracil to cytosine takes place in the triphosphate form, catalyzed by CTP synthetase. Glutamine is the nitrogen donor, and ATP is required, as we saw earlier in similar reactions.

$$UTP + \text{glutamine} + ATP \rightarrow CTP + \text{glutamate} + ADP + Pi$$

Feedback inhibition in pyrimidine nucleotide biosynthesis takes place in several ways. CTP is an inhibitor of aspartate transcarbamoylase and of CTP synthetase. UMP is an inhibitor of an even earlier step, the one catalyzed by carbamoyl phosphate synthetase.

Pyrimidine nucleotides are broken down first to the nucleoside and then to the base, as purine nucleotides are. Cytosine can be deaminated to uracil, and the double bond of the uracil ring is reduced to produce dihydrouracil. The ring opens to produce N-carbamoylpropionate, which in turn is broken down to NH_4, CO_2, and 13-alanine.

Reductiopn of ribonucieotides

Ribonucleoside diphosphates are reduced to 2'-deoxyribonucleoside diphosphates in all organisms; NADPH is the reducing agent.

$$\text{Ribonucleoside diphosphate} + \text{NADPH} + \text{H}^+ \rightarrow \text{deoxyribonucleoside diphosphate} + \text{NADP}^+ + \text{H}_2\text{O}$$

The actual process, which is catalyzed by ribonucleotide reductase, is more complex than the preceding equation would indicate and involves some intermediate electron carriers. The ribonucleotide reductase system from E. coli has been extensively studied, and its mode of action gives some clues to the nature of the process. Two other proteins are required, thioredoxin and thioredoxin reductase.

Thioredoxin contains a disulfide (S-S) group in its oxidized form and two sulfhydryl (-SH) groups in its reduced form. NADPH reduces thioredoxin in a reaction catalyzed by thioredoxin reductase. The reduced thioredoxin in turn reduces a ribonucleoside diphosphate (NDP) to a deoxyribonucleoside diphosphate (dNDP), and it is this reaction that is actually catalyzed by ribonucleotide reductase. This reaction produces dADP, dGDP, dCDP, and dUDP. The first three are phosphorylated to give the corresponding triphosphates, which are substrates for the synthesis of DNA. Another required substrate for DNA synthesis is dTTP, and we shall now see how dTTP is produced from dUDP.

Conversion of dUDP and to dTTP

A one-carbon transfer is required for the conversion of uracil to thymine by attachment of the methyl group. The most important reaction in this conversion is that catalyzed by thymidylate synthetase. The source of the one-carbon unit is N^5, N^{10}-methylenetetrahydrofolate, which is converted to dihydrofolate in the process. The metabolically active form of the one-carbon carrier is tetrahydrofolate.

Dihydrofolate must be reduced to tetrahydrofolate for this series of reactions to continue, and this process requires NADPH and dihydrofolate reductase. Since a supply of dTTP is necessary for DNA synthesis, inhibition of enzymes that catalyze the production of dT]FP will inhibit the growth of rapidly dividing cells.

Cancer cells, like all fast-growing cells, depend on continued DNA synthesis for growth. Inhibitors of thymidylate synthetase, such as fluorouracil, and inhibitors of dihydrofolate reductase, such as aminopterin and methotrexate (structural analogues of folate), have been used in cancer chemotherapy.

The intent of such therapy is to inhibit the formation of dTTP and thus of DNA in cancer cells, causing the death of the cancer cells with minimal effect on normal cells, which grow more slowly. Chemotherapy has adverse side effects because of the highly toxic nature of most of the drugs involved; normal cells are affected to some extent, although less than the cancer cells. Enormous amounts of research are focused on finding safe and effective forms of treatment.

The metabolism of nitrogen encompasses a number of topics, including the anabolism and catabolism of amino acids, porphyrins, and nucleotides. Atmospheric nitrogen is the ultimate source of this element in biomolecules. Nitrogen fixation is the process by which molecular nitrogen from the atmosphere is made available to organisms, in the form of ammonia. Nitrification reactions convert NO_3 to NH_3 and provide another source of nitrogen. Feedback control mechanisms are a unifying factor in biosynthetic pathways involving nitrogen compounds.

In the anabolism of amino acids, transamination reactions play an important role. Glutamate and glutamine are frequently the amino group donors. The enzymes that catalyze transamination reactions frequently require pyridoxal phosphate as a coenzyme.

One-carbon transfers also operate in the anabolism of amino acids. Carriers are required for the one-carbon groups transferred. The trahydrofolate is a carrier of methylene and formyl groups, and S-adenosylmethionine is a carrier of methyl groups. Some species, including humans, cannot synthesize all the amino acids required for protein synthesis and must obtain these essential amino acids from dietary sources. The catabolism of amino acids has two parts: the fate of the nitrogen and the fate of the carbon skeleton.

In the urea cycle, nitrogen released by the catabolism of amino acids is converted to urea. The carbon skeleton is converted to pyruvate or oxaloacetate in the case of glucogenic amino acids, or to acetyl-CoA or acetoacetyl-CoA in the case of ketogenic amino acids.

Chapter 7

Lipid Metabolism

The definition of a lipid is based on solubility. Lipids are marginally soluble in water and soluble in organic solvents such as chloroform or acetone. Fats and oils are typical lipids in terms of their solubility, but that fact does not really define their chemical nature. In terms of chemistry, lipids are a mixed bag of compounds that share some properties based on structural similarities, mainly a preponderance of nonpolar groups.

Classified according to their chemical nature, the lipids fall into two main groups. One group, which consists of open-chain compounds with polar head groups and long nonpolar tails, includes fatty acids, triacylglycerols, sphingolipids, phosphoacylglycerols, and glycolipids. The second major group consists of fused-ring compounds, the steroids; an important representative of this group is cholesterol.

Fatty acids

Fatty acids have a carboxyl group at the polar end and a hydrocarbon chain at the nonpolar tail. Fatty acids are amphipathic compounds because the carboxyl group is hydrophilic and the hydrocarbon tail is hydrophobic. The carboxyl group can ionize under the proper conditions. A fatty acid that occurs in a living system normally contains an even number of carbon atoms, and the hydrocarbon chain is usually unbranched. If there are carbon-carbon double bonds in the chain, the fatty acid is unsaturated; if there are only single bonds, the fatty acid is saturated. In unsaturated fatty acids the stereochemistry at the double bond is usually *cis* rather than *trans*.

Note that the double bonds are isolated from one another by several singly bonded carbons; fatty acids do not normally have conjugated double bond systems. The notation used for fatty acids indicates the number of carbon atoms and the number of double bonds. In this system 18:0 denotes an 18-carbon saturated fatty acid (no double bonds), and 18:1 denotes an 18-carbon fatty acid with one double bond.

In the unsaturated fatty acids there is a double bond at the ninth carbon atom from the carboxyl end. The position of the double bond results from the way unsaturated fatty acids are synthesized in organisms.

Unsaturated fatty acids have lower melting points than saturated ones. Plant oils are liquid at room temperature because they have higher proportions of unsaturated fatty acids than do animal fats, which tend to be solids. Conversion of oils to fats is a commercially important process. It involves hydrogenation, the process of adding hydrogen across the double bond of unsaturated fatty acids to produce the saturated counterpart. Oleomargarine manufacture in particular makes extensive use of partially hydrogenated vegetable oils. Fatty acids are rarely found free in nature, but they form parts of many commonly occurring lipids.

Triacylglycerols

When all three of the alcohol groups form ester linkages with fatty acids, the resulting compound is a triacylglycerol; an older name for this type of compound is triglyceride. Note that the three ester groups are the polar part of the molecule, whereas the tails of the fatty acids are nonpolar. It is usual for three different fatty acids to be esterified to the alcohol groups of the same glycerol molecule.

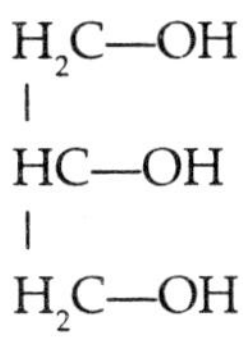

Glycerol

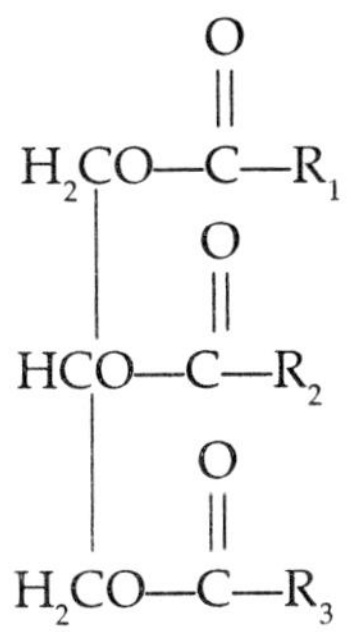

Triacylglycerol

Fig. 1. Structure of glicerol and a triacylglycerol

Triacylglycerols do not occur as components of membranes (as do other types of lipids), but they accumulate in adipose tissue (primarily fat cells) and provide a means of storing fatty acids, particularly in animals. They serve as concentrated stores of metabolic energy; complete oxidation of fats yields about 9 kcal/g, in contrast to 4 kcal/g for carbohydrates and proteins. When fatty acids are put to use by an organism, the ester linkages of triacylglycerols are hydrolyzed by enzymes called lipases.

The same hydrolysis take place outside organisms, with acids or bases as catalysts reaction can when a base such as sodium hydroxide or potassium hydroxide is used, the products of the reaction, which is called saponification, are glycerol and the sodium or potassium salts of the fatty acids.

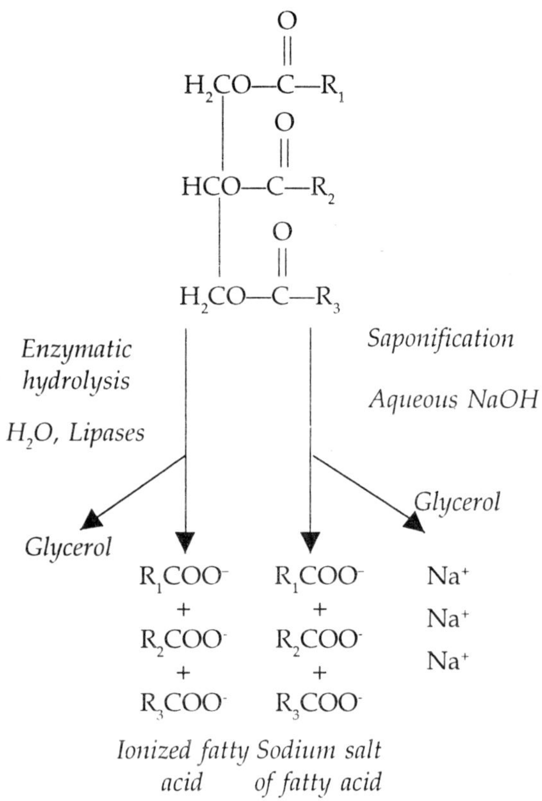

Fig. 2. Hydrolysis of triacylglycerols

When soaps are used with hard water, the calcium and magnesium ions in the water react with the fatty acids and form a precipitate—the characteristic scum left on the insides of sinks and bathtubs. The other product of saponification, glycerol, is used in creams and lotions as well as for the manufacture of nitroglycerin.

Phosphoacylglycerols

One of the alcohol groups of glycerol to be esterified by a phosphoric acid molecule rather than by a carboxylic acid.

(a)

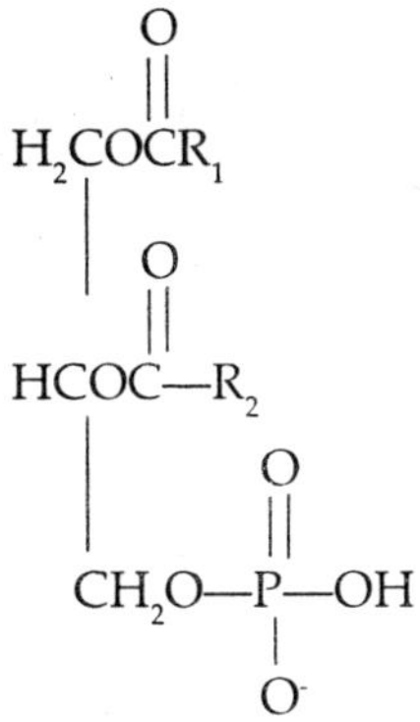

Phosphatidic acid

(b)

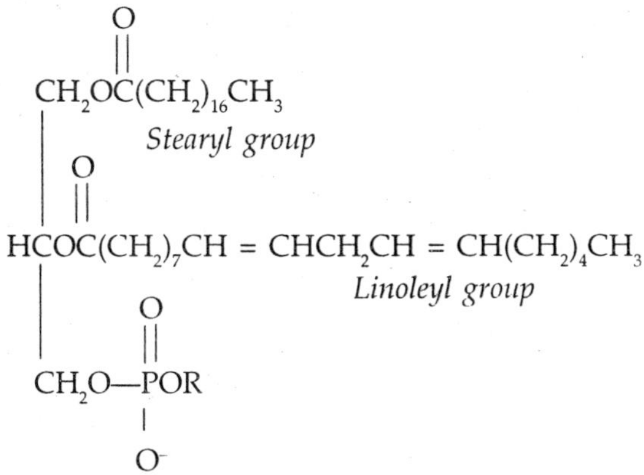

Phosphatidyl ester

Fig. 3. The molecular architecture of phosphoacylglycerols. (a) A phosphatidic acid, in which glycerol is esterified to phosphoric acid and to two different carboxylic acids. (b) A phosphatidyl

In such lipid molecules, two fatty acids are also esterified to the glycerol molecule. The resulting compound is called a phosphatidic acid. Fatty acids are usually monoprotic acids with only one carboxyl group able to form an ester bond, but phosphoric acid is triprotic and thus can form more than one ester linkage. One molecule of phosphoric acid can form ester bonds both to glycerol and to some other alcohol, creating a phosphatidyl ester.

Phosphatidyl esters are classed as phosphoacylglycerols. The natures of the fatty acids vary widely, as they do in triacylglycerols. As a result, the names of the types of lipids (such as triacylglycerols and phosphoacylglycerols) that contain fatty acids must be considered generic names. The classification of a phosphatidyl ester depends on the nature of the second alcohol esterified to the phosphoric acid.

Some of the most important lipids in this class are phosphatidyl ethanolamine (cephalin), phosphatidyl serine, phosphatidyl choline (lecithin), phosphatidyl inositol, phosphatidyl glycerol, and diphosphatidyl glycerol (cardiolipin). In each of these types of compounds the natures of the fatty acids in the molecule can vary widely.

All these compounds have long, nonpolar, hydrophobic tails and polar, highly hydrophilic head groups and thus are markedly amphipathic. In a phosphoacylglycerol the polar head group is charged, since the phosphate group is ionized at neutral pH. There is frequently also a positively charged amino group contributed by an amino alcohol esterified to the phosphoric acid. Phosphoacylglycerols are important components of biological membranes.

Waxes

Waxes are complex mixtures of esters of long-chain carboxylic acids and long-chain alcohols. They frequently serve as protective coatings for both plants and animals. In plants, they coat stems, leaves, and fruit; in animals, they are found on fur, feathers, and skin.

$$CH_3—(CH_2)_{24}—\overset{\overset{\large O}{\|}}{C}—O—(CH_2)_{29}—CH_3$$

Myricyl cerotate

$$CH_3—(CH_2)_{14}—\overset{\overset{\large O}{\|}}{C}—O—(CH_2)_{15}—CH_3$$

Cetyl palmitate

Fig. 4. Structures of some waxes and sphingolipids.

Myricyl cerotate, the principal component of carnauba wax, is produced by the Brazilian wax palm. Carnauba wax is extensively used in floor wax and automobile wax. The principal component of spermaceti, a wax produced by whales, is cetyl palmitate. The use of spermaceti as a fragrance made it one of the most highly prized products of 19th-century whaling efforts.

Sphingolipids

Sphingolipids are found in both plants and animals; they are particularly abundant in the nervous system. The simplest compounds of this class are the ceramides, which consist of one fatty acid linked to the amino group of sphingosine by an amide bond. In sphingomyelins the primary alcohol group of sphingosine is esterified to phosphoric acid, which in turn is esterified to another amino alcohol, choline.

Note the structural similarities between sphingomyelin and other phospholipids. Two long hydrocarbon chains are attached to a backbone that contains alcohol groups. One of the alcohol groups of the backbone is esterified to phosphoric acid. A second alcohol, choline in this case, is also esterified to the phosphoric acid. We have already seen that choline occurs in phosphoacylglycerols. Sphingomyelins are amphipathic; they occur in cell membranes in the nervous system.

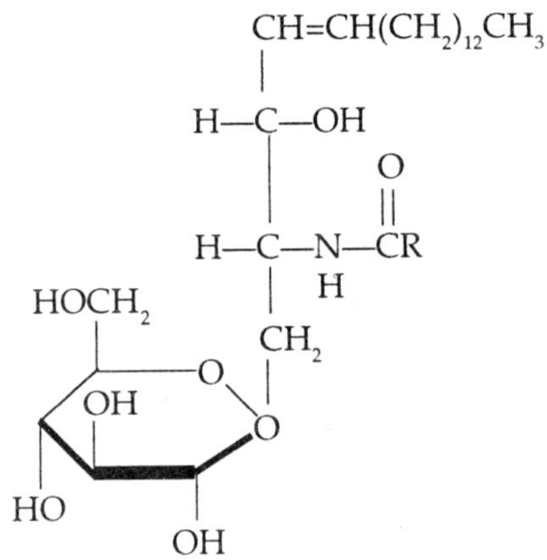

A Glucocerebroside

Glycolipids

Ceramides are the parent compounds for glycolipids, and the glycosidic bond is formed between the primary alcohol group of the ceramide and a sugar residue. The resulting compound is called a cerebroside. In most cases the sugar is glucose or galactose; for example, a glucocerebroside is a cerebroside that contains glucose. As the name indicates, cerebrosides are found in nerve and brain cells, primarily in cell membranes. The carbohydrate portion of these compounds can be very complex. Gangliosides, are examples of glycolipids with a complex carbohydrate moiety.

Steroids

Many compounds of widely differing functions are classified as steroids on the basis of the same general structure: a fused-ring system consisting of three six-membered rings (the A, B, and C rings) and one five-membered ring (the D ring). There are many important steroids, including sex hormones. The steroid that is of most interest in our discussion of membranes is cholesterol. The only hydrophilic group in the cholesterol structure is the single hydroxyl group. As a result, the molecule is highly hydrophobic.

Cholesterol is widespread in biological membranes, especially in animals, but it does not occur in prokaryotic cell membranes. In spite of its many important biological functions, including its role as a precursor of other steroids and of vitamin D_3, cholesterol is best known for its harmful effects on health. It plays a role, in the development of atherosclerosis, a condition in which lipid deposits block the blood vessels and lead to heart disease.

Biological membranes

The molecular basis of the membrane's structure lies in its lipid and protein components. Now it is time to see how the interaction between the lipid bilayer and membrane proteins determines membrane function. Membranes not only separate cells from the external environment they also play important roles in transport of specific substances into and out of cells. In addition, a number of important enzymes are found in membranes and depend on this environment for their function. Phosphoglycerides are prime examples of amphipathic molecules, and they are the principal lipid components of membranes. The existence of lipid bilayers depends on hydrophobic interactions. These bilayers, are frequently used as models for biological membranes because they have many features in common with them—such as a hydrophobic interior and an ability to control the transport of small molecules and ions -yet are simpler and easier to work with in the laboratory. The most important difference between lipid bilayers and cell membranes is that the latter contain proteins as well as lipids. The protein component of a membrane can make up 20% to 80% of its total weight. An understanding of membrane structure requires knowledge of how the protein and lipid components contribute to the properties of the membrane.

Lipid bilayers

Steroids are present in eukaryotes-cholesterol in animal membranes and similar compounds called phytosterols in plants. In the lipid-bilayer part of the membrane, the polar head groups are in contact with water, and the nonpolar tails lie in the interior. The whole bilayer arrangement is held together by noncovalent interactions such as van der Waals and hydrophobic interactions. The surface of the bilayer is polar and contains charged groups.

The nonpolar hydrocarbon interior of the bilayer consists of the saturated and unsaturated chains of fatty acids and the fused ring system of cholesterol. Both the inner and outer layers of the bilayer contain mixtures of lipids, but their compositions differ and can be used to distinguish the inner and outer layers from each other. Bulkier molecules tend to occur in the outer layer, and smaller molecules tend to occur in the inner layer. The arrangement of the hydrocarbon interior of the bilayer can be ordered and rigid or disordered and fluid. The bilayer's fluidity depends on its composition. In saturated fatty acids, a linear arrangement of the hydrocarbon chains leads to close packing of the molecules in the bilayer, and thus to rigidity. In unsaturated fatty acids there is a kink in the hydrocarbon chain that does not exist in saturated fatty acids.

The kinks cause disorder in the packing of the chains, with a more open structure than would be possible for straight saturated chains. In turn, the disordered structure caused by the presence of unsaturated fatty acids with cis double bonds and, thus, kinks in their hydrocarbon chains causes greater fluidity in the bilayer. The lipid components of a bilayer are always in motion, to a greater extent in more fluid bilayers and to a lesser extent in more rigid ones. The presence of cholesterol may also enhance order and rigidity. The fused-ring structure of cholesterol is itself quite rigid, and the presence of cholesterol stabilizes the extended straight-chain arrangement of saturated fatty acids by van der Waals interactions.

The lipid portion of a plant membrane has a higher percentage of unsaturated fatty acids, especially polyunsaturated (containing two or more double bonds) fatty acids, than the lipid portion of an animal membrane. Furthermore, the presence of cholesterol is characteristic of animal, rather than plant, membranes; other steroids (phytosterols) are found in plant membranes. As a result, animal membranes are less fluid (more rigid) than plant membranes, and the membranes of prokaryotes, which contain no appreciable amounts of steroids, are the most fluid of all. With heat, ordered bilayers become less ordered; bilayers that are comparatively disordered become even more disordered. This is a cooperative transition that takes place at a characteristic temperature, like the melting of a crystal (which is also a cooperative transition).

The transition temperature is higher for more rigid and ordered membranes than it is for relatively fluid and disordered membranes. Some connections between fatty acid composition of bilayers and membranes and how they behave at different temperatures are discussed in the box. Recall that the distributions of lipids are not the same in the inner and outer portions of the bilayer. Since the bilayer is curved, the molecules of the inner layer are more tightly packed. Bulkier molecules such as cerebrosides tend to be located in the outer layer. There is very little tendency for "flip-flop" migration of lipid molecules from one layer of the bilayer to another. Lateral motion of lipid molecules within one of the two layers frequently takes place, however, especially in more fluid bilayers.

Several methods exist for monitoring the motions of molecules within a lipid bilayer. These methods depend on labelling some part of the lipid component with an easily detected "tag." The tags are usually fluorescent compounds, which can be detected with high sensitivity. Another kind of labelling method depends on the fact that some nitrogen compounds have unpaired electrons. These compounds are used as labels and can be detected by magnetic measurements.

Lipids and membrane proteins

Proteins in a biological membrane can be associated with the lipid bilayer in either of two ways as peripheral proteins on the surface of the membrane or as integral proteins within the lipid bilayer. Peripheral proteins are usually bound to the charged head groups of the lipid bilayer by polar or electrostatic interactions, or both. They can be removed by such mild treatment as raising the ionic strength of the medium. The relatively numerous charged particles present in a medium of higher ionic strength undergo more electrostatic interactions with the lipid and with the protein, "swamping out" the comparatively fewer electrostatic interactions between the protein and the lipid.

It is much more difficult to remove integral proteins from membranes. Harsh conditions, such as treatment with detergents or extensive sonication (exposure to ultrasonic vibrations), are usually required. Such measures frequently denature the protein, which often remains bound to lipids in spite of all efforts to obtain it in pure form.

The denatured protein is of course inactive, whether or not it remains bound to lipids. Fortunately, nuclear magnetic resonance techniques are now enabling researchers to study proteins of this sort in living tissue. The structural integrity of the whole membrane system appears to be necessary for the activities of most membrane proteins. Membrane proteins have a variety of functions.

Most, but not all, of the important functions of the membrane as a whole are those of the protein component. Transport proteins help move substances in and out of the cell, and receptor proteins are important in the uptake of extracellular signals, such as hormones or neurotransmitters, into the cell. In addition, some enzymes are tightly bound to membranes; examples include many of the enzymes responsible for aerobic oxidation reactions, which are found in specific parts of mitochondrial membranes.

Some of these enzymes are oil the inner surface of the membrane, and some are on the outer surface. There is an uneven distribution of proteins of all types on the inner and outer layers of all cell membranes, just as there is an asymmetric distribution of lipids.

We have seen that biological membranes have both lipid and protein components. How do these two parts combine to produce a biological membrane? The fluid mosaic model is the most widely accepted description of biological membranes at the moment. The term "mosaic" implies that the two components exist side by side without forming some other substance of intermediate nature.

No extensive formation of lipid-protein complexes takes place, for example. Instead, the basic structure of biological membranes is that of the lipid bilayer, with the proteins embedded in the bilayer structure. These proteins tend to have a specific orientation in the membrane. The term "fluid mosaic" implies that the same sort of lateral motion occurs in membranes that we have already seen in lipid bilayers. The proteins "float" in the lipid bilayer and can move along the plane of the membrane.

Electron micrographs can be made of membranes that have been frozen and then fractured along the interface between the two layers. The outer layer is removed, exposing the interior of the membrane. The interior has a granular appearance because of the presence of the integral membrane proteins.

Three important functions take place in or on inembranes (in addition to the structural role of membranes as the boundaries and containers of all cells and of the organelles within eukaryotic cells). The first of these functions is transport. Membranes are semipermeable barriers to the flow of substances into and out of cells and organelles. Transport through the membrane can involve the lipid bilayer as well as the membrane proteins. The other two important functions primarily involve the membrane proteins. One of these functions is catalysis.

Enzymes can be bound-in some cases very tightly - to membranes, and the enzymatic reaction takes place on the membrane. The third significant function is the receptor property, in which proteins bind specific biologically important substances that trigger biochemical responses in the cell.

The most important question about transport of substances across biological membranes is whether or riot the process requires expenditure of energy by the cell. In passive transport a substance moves from a region of higher concentration to one of lower concentration. In other words, the movement of the substance is in the same direction as a concentration gradient, and no energy is expended by the cell.

In active transport a substance moves from a region of lower concentration to one of higher concentration (against a concentration gradient), and this process does require expenditure of energy by the cell.

The process of passive transport can be subdivided into two categories, simple diffusion and facilitated diffusion. In simple diffusion a molecule moves directly through an opening or pore in the membrane without interacting with another molecule. The transported molecule goes either directly through the lipid bilayer or through an opening in a channel protein.

Small molecules such as water, oxygen, and carbon dioxide can pass directly through the lipid bilayer, but larger molecules, especially polar ones, require a channel protein. Ions cannot pass through the lipid bilayer because of their charge, and they also require channel proteins. The sizes and shapes of the openings in channel proteins make them specific for given ions or molecules.

In facilitated diffusion the molecule to be transported across the membrane binds to a carrier protein rather than simply passing through, as it would a channel protein in simple diffusion. In both a channel protein and a carrier protein a pore is created by the folding of the backbone and side chains. Many of these proteins have several α-helical portions that span the membrane; in others a β-barrel forms the pore. In one example the helical portion of the protein spans the membrane.

The exterior, which is in contact with the lipid bilayer, is hydrophobic, whereas the interior, through which ions pass, is hydrophilic. Note that this orientation is the inverse of that observed in water-soluble globular proteins.

Active transport requires moving substances against a concentration gradient. The situation is so markedly similar to pumping water uphill that one of the most extensively studied examples of active transport, moving potassium ions into a cell and simultaneously moving sodium ions out of the cell, is referred to as the sodium-potassium ion pump.

Under normal circumstances the concentration of K^+ is higher inside a cell than in extracellular fluids ($[K^+]_{inside} > [K^+]_{outside}$), but the concentration of Na^+ is lower inside the cell than out ($[Na^+]_{inside} < [Na^+]_{outside}$). The energy required to move these ions against their gradients comes from an exergonic (energy-releasing) reaction, the hydrolysis of ATP to ADP and Pi (phosphate ion). There can be no transport of ions without hydrolysis of ATP. The same protein appears to serve as both the enzyme that hydrolyzes the ATP (the ATPase) and the transport protein; it consists of several subunits. The reactants and products of this hydrolysis reaction - ATP, ADP, and phosphate ion - remain within the cell, and the phosphate becomes covalently bonded to the transport protein for part of the process. The Na^+-K^+ pump operates in several steps.

One subunit of the protein hydrolyzes the ATP and transfers the phosphate group to an aspartate side chain on another subunit. The bond formed here is a mixed anhydride. Simultaneously, binding of 3 Na^+ from the interior of the cell takes place.

The phosphorylation of one subunit causes a conformational change in the protein, which opens a channel or pore through which the 3 Na^+ can be released to the extracellular fluid. Outside the cell, $2K^+$ bind to the pump enzyme, which is still phosphorylated.

Another conformational change occurs when the bond between the enzyme and the phosphate group is hydrolyzed. This second conformational change regenerates the original form of the enzyme and allows the $2K^+$ to enter the cell.

The pumping process transports three Na^+ ions out of the cell for every two K^+ ions transported into the cell. The operation of the pump can be reversed when there is no K^+ and a high concentration of Na^+ in the extracellular medium; in this case, ATP is produced by the phosphorylation of ADP.

The actual operation of the Na^+-K^+ pump is not completely understood and probably is even more complicated than we now know. There is also a calcium ion (Ca^{2+}) pump, which is a subject of equally active investigation. Unanswered questions about the detailed mechanism of active transport provide opportunities for future research.

The interaction between receptor proteins and the active substances to which they bind has features in common with enzyme-substrate recognition. There is a requirement for essential functional groups that have the right three-dimensional conformation with respect to one another.

The binding site, whether on a receptor or an enzyme, must provide a good fit for the substrate. In receptor binding, as in enzyme behaviour, there -is a possibility for inhibition of the action of the protein by some sort of "poison" or inhibitor. The study of receptor proteins is less advanced than the study of enzymes, since many receptors are tightly bound integral proteins and their activity depends on the membrane environment.

Receptors are often large oligomeric proteins (ones with several subunits), with molecular weights on the order of hundreds of thousands. Also, quite frequently there are very few molecules of the receptor in each cell, adding to the difficulties of isolating and studying this type of protein.

An important type of receptor is that for low-density lipoprotein (LDL), the principal carrier of cholesterol in the bloodstream. LDL is a particle that consists of various lipids-in particular, cholesterol and phosphoglycerides as well as a protein. The protein portion of the LDL particle binds to the LDL receptor of a cell. The complex formed between the LDL and the receptor is pinched off into the cell in a process called endocytosis.

The receptor protein is then recycled back to the surface of the cell. The cholesterol portion of the LDL is used in the cell, but an oversupply of cholesterol causes problems. Excess of cholesterol inhibits the synthesis of LDL receptor. If there are too few receptors for LDL, the level of cholesterol in the bloodstream increases. Eventually the excess cholesterol is deposited in the arteries, blocking them severely. This blocking of arteries, called atherosclerosis, can eventually lead to heart attacks and strokes.

Lipid-soluble vitamins

Some vitamins, having a variety of functions, are of interest in this chapter because they are soluble in lipids. They are themselves hydrophobic, which accounts for their solubility.

Vitamin A

The extensively unsaturated hydrocarbon β-carotene is the precursor of vitamin A, which is also known as retinol. As the name suggests, β-carotene is abundant in carrots, and it also occurs in other vegetables, particularly the yellow ones. When an organism requires vitamin A, β-carotene is converted to the vitamin.

A derivative of vitamin A plays a crucial role in vision when it is bound to a protein call opsin. The cone cells in the retina of the eye contain several types of opsin and are responsible for vision in bright light and for colour vision. The rod cells in the retina contain only one type of opsin; they are responsible for vision in dim light. The chemistry of vision has been more extensively studied in rod cells than in cone cells, and we shall discuss events that take place in rod cells.

Vitamin A has an alcohol group that is enzymatically oxidized to an aldehyde group, forming retinal. Two isomeric forms of retinal, involving cis-trans isomerization around one of the double bonds, are important in the behaviour of this compound in vivo. The aldehyde group of retinal forms an imine (also called a Schiff base) with the side-chain amino group of a lysine residue in rod-cell opsin. The product of the reaction between retinal and opsin is rhodopsin. The outer segment of rod cells contains flat membrane-bounded discs, the membrane consisting of about 60% rhodopsin and 40% lipid.

Table 1. Lipid-soluble vitamins and their functions

Vitamin	*Function*
Vitamin A	Servers as the site of the primary photochemical reaction in vision
Vitamin D	Regulates calcium (and phosphorus) metabolism
Vitamin E	Serves as an antioxidant; necessary for reproduction in rats and may be necessary for reproduction in humans
Vitamin K	Has a regulatory function in blood cclotting

Vitamin D

The several forms of vitamin D play a major role in regulation of calcium and phosphorus metabolism. One of the most important of these compounds, vitamin D_3 (cholecalciferol), is formed from cholesterol by the action of ultraviolet radiation from the sun. Vitamin D_3 is further processed in the body to form hydroxylated derivatives, which are the metabolically active form of this vitamin.

The presence of vitamin D_3 leads to increased synthesis of a Ca^{2+}-binding protein, which increases absorption of dietary calcium in the intestines. This process results in calcium uptake by the bones. A deficiency of vitamin D can lead to rickets, a condition in which the bones of growing children become soft, resulting in skeletal deformities. Children, especially infants, have higher requirements for vitamin D than do adults. Milk with vitamin D supplements is available to most children. Adults who are exposed to normal amounts of sunlight do not usually require vitamin D supplements.

Vitamin E

The most active form of vitamin E is α-tocopherol. In rats, vitamin E is required for reproduction and for prevention of the disease muscular dystrophy. It is not definitely known whether this requirement exists in humans.

A well-established chemical property of vitamin E is that it is an antioxidant that is, a good reducing agent - so it reacts with oxidizing agents before they can attack other biomolecules. The antioxidant action of vitamin E has been shown to protect important compounds, including vitamin A, from degradation in the laboratory; it probably, also serves this function in organisms.

Recent research has shown that the interaction of vitamin E with membranes enhances its effectiveness as an antioxidant. Another function of antioxidants such as vitamin E is to react with, and thus remove, the very reactive and highly dangerous substances known as free radicals. A free radical has at least one unpaired electron, which accounts for its high degree of reactivity. Free radicals may play a part in the development of cancer and in the aging process.

Vitamin K

The name of vitamin K comes from the German Koagulation, because this vitamin is an important factor in the blood clotting process. The bicyclic ring system contains two carbonyl groups, the only polar groups on the molecule.

There is a long unsaturated hydrocarbon side chain that consists of repeating isoprene units, the number of which determines the exact form of vitamin K Several forms of this vitamin can be found in a single organism, and the reason for this variation is not well understood. Vitamin K is not the first vitamin we have encountered that contains isoprene units, but it is the first one in which the number of isoprene units and their degree of saturation makes a difference. It is also known that the steroids are biosynthetically derived from isoprene units, but the structural relationship is not immediately obvious.

The presence of vitamin K is required in the complex process of blood clotting, which involves many steps and many proteins and has stimulated numerous unanswered questions. It is known definitely that vitamin K is required to modify prothrombin and other proteins involved in the clotting process.

Specifically, with prothrombin the side chains of several glutamate residues of prothrombin are altered by the addition of another carboxyl group. This modification of glutamate produces γ-carboxylglutamate residues. The two carboxyl groups in proximity form a bidentate ("two teeth") ligand, which can bind calcium ion (Ca^{2+}). If prothrombin is not modified in this way, it does not bind Ca^{2+}.

Even though there is a lot more to be learned about blood clotting and the role of vitamin K in the process, this point at least is well established, because Ca^{2+}, is required for blood clotting. Parenthetically, two well-known anticoagulants, dicumarol and warfarin (a rat poison), are vitamin K antagonists.

Prostaglandins and Leukotrienes

A group of compounds derived from fatty acids has a wide range of physiological activities; they are called prostaglandins because they were first detected in seminal fluid, which is produced by the prostate gland. It has since been shown that they are widely distributed in a variety of tissues.

The metabolic precursor of all prostaglandins is arachidonic, a fatty acid that contains 20 carbon atoms and four double bonds. The double bonds are not conjugated. The production of the prostaglandins from arachidonic acid takes place in several steps, which are catalyzed by enzymes.

The prostaglandins themselves each have a five-membered ring; they differ from one another in the numbers and positions of double bonds and oxygen containing functional groups.

Elucidation of the structures of prostaglandins and their laboratory synthesis have been topics of great interest to organic chemists, largely because of the many physiological effects of these compounds and their possible usefulness in the pharmaceutical industry. Some of the functions of prostaglandins are control of blood pressure, stimulation of smooth muscle contraction, and induction of inflammation.

Aspirin inhibits the synthesis of prostaglandins, particularly in blood platelets, a property that accounts for its antiinflammatory and fever-reducing properties. Cortisone and other steroids also have antiinflammatory effects because of their inhibition of prostaglandin synthesis.

Prostaglandins are known to inhibit the aggregation of platelets. They may thus be of therapeutic value by preventing the formation of blood clots, which can cut off the blood supply to the brain or the heart and cause certain types of strokes and heart attacks. Even if this behaviour were the only useful property of prostaglandins, it would justify considerable research effort.

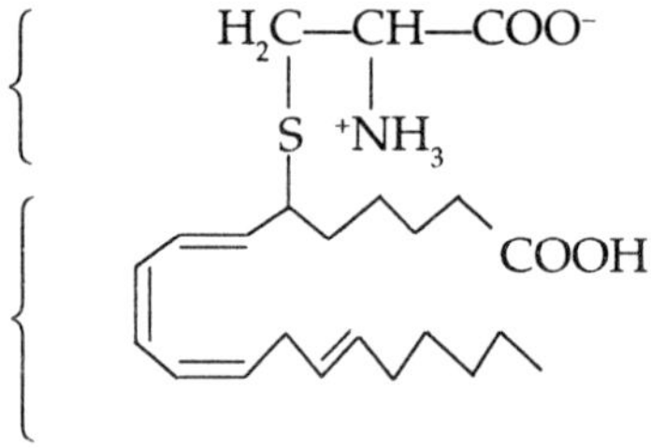

Fig. 5. Leukotriene C.

Heart attacks and strokes are two of the leading causes of death in industrialized countries. More recently, the study of prostaglandins has been a topic of great interest because of their possible antitumor and antiviral activity.Leukotrienes are compounds that, like prostaglandins, are derived from arachidonic acid. They are found in leukocytes (white blood cells) and have three conjugated double bonds; these two facts account for their name. (Fatty acids and their derivatives do not normally contain conjugated double bonds.)

Leukotriene C is a typical member of this group; note the 20 carbon atoms in the carboxylic acid backbone, a feature that relates this compound structurally to arachidonic acid. (The 20-carbon prostaglandins and leukotrienes are also called eicosinoids.) An important property of leukotrienes is their constriction of smooth muscle, especially in the lungs. Asthma attacks may result from this constricting action, since the synthesis of leukotriene C appears to be facilitated by allergic reactions, such as a reaction to pollen. Drugs that inhibit the synthesis of leukotriene C are now being used in the treatment of asthma. In addition, other drugs are designed to block leukotriene receptors.

Thromboxanes are a third class of derivatives of arachidonic acid. They contain cyclic ethers as part of their structures. The most widely studied member of the group, Thromboxane A_2 (TxA_2), is known to induce platelet aggregation and smooth muscle contraction.

Catabolism of lipids

Oxidation of fatty acids is the chief source of energy in the catabolism of lipids; in fact, lipids that are sterols are not catabolized as a source of energy but are excreted. Both triacylglycerols, which are the main storage form of the chemical energy of lipids, and phosphoacylglycerols, which are important components of biological membranes, have fatty acids as part of their covalently bonded structures. In both types of compounds, the bond between the fatty acid and the rest of the molecule can be hydrolyzed, with the reaction catalyzed by suitable groups of enzymes, lipases in the case of triacylglycerols and phospholipases in the case of phosphoacylglycerols.

Several different phospholipases can be distinguished on the basis of site at which they hydrolyze phospholipids. Phospholipase A_2 is widely distributed in nature; it is also being actively studied by biochemists interested in its structure and mode of action, which involves hydrolysis of phospholipids at the surface of micelles.

The reason for the name phospholipase A_2 rather than phospholipase B is found in a bit of biochemical history. At one time it was thought that phospholipase A_2 hydrolyzed the fatty acid at the A position. Later biochemists found that the point of attack was at the B-position, but they kept the name phospholipase A with the added subscript.

Phospholipase D occurs in spider venom and is responsible for the tissue damage that accompanies spider bites. Snake venoms also contain phospholipases; the concentration of phospholipases is particularly high in venoms with comparatively low concentrations of the toxins that are characteristic of some kinds of venom. The lipid products of hydrolysislyse red blood cells, preventing clot formation. Snakebite victims bleed to death in this situation.

Fatty acid oxidation begins with activation of the molecule. In this reaction a thioester bond is formed between the carboxyl group of the fatty acid and the thiol group of coenzyme A (CoA-SH). The activated form of the fatty acid is an acyl-CoA, the exact nature of which depends on the nature of the fatty acid itself. Keep in mind throughout this discussion that all acyl-CoA molecules are thioesters, since the fatty acid is esterified to the thiol group of CoA. The enzyme that catalyzes formation of the ester bond, an acyl-CoA synthetase, requires ATP for its action.

In the course of the reaction, an acyl adenylate intermediate is formed. The acyl group is then transferred to CoA-SH. ATP is converted to AMP and PPi, rather than to ADP and Pi. The PPi is hydrolyzed to two Pi; the hydrolysis of two "high-energy" phosphate bonds provides energy for the activation of the fatty acid and is equivalent to the use of two ATP. The formation of the acyl-CoA is endergonic without the energy provided by the hydrolysis of the two "high-energy" bonds. Note also that the hydrolysis of ATP to AMP and two Pi represents an increase in entropy.

There are several enzymes of this type, some specific for longer-chain and some for shorter-chain fatty acids. Both saturated and unsaturated fatty acids can serve as substrates for these enzymes. The esterification takes place in the cytosol, but the rest of the reactions of fatty acid oxidation occur in the mitochondrial matrix. The activated fatty acid must be transported into the mitochondrion so that the rest of the oxidation process can proceed.

The acyl-CoA can cross the outer mitochondrial membrane but not the inner membrane. In the intermembrane space the acyl group is transferred to carnitine by transesterification; this reaction is catalyzed by the enzyme carnitine acyltransferase, which is located in the inner membrane. Acyl carnitine, a compound that can cross the inner mitochondrial membrane, is formed. In the matrix the acyl group is transferred from carnitine to mitochondrial CoA-SH by another transesterification reaction.

In the matrix a repeated sequence of reactions successively cleaves two-carbon units from the fatty acid, starting from the carboxyl end. This process is called P-oxidation, since the oxidative cleavage takes place at the P-carbon of the acyl group esterified to CoA. The B-carbon of the original fatty acid becomes the carboxyl carbon in the next stage of degradation. The whole cycle requires four reactions.

1. The acyl-CoA is oxidized to an α, β unsaturated acyl-CoA . The product has the trans arrangement at the double bond. This reaction is catalyzed by an FAD-dependent acyl-CoA dehydrogenase.

2. The unsaturated acyl-CoA is hydrated to produce α, β-hydroxyacyl-CoA. This reaction is catalyzed by the enzyme enoyl-CoA hydratase.

3. A second oxidation reaction is catalyzed by B-hydroxyacyl-CoA dehydrogenase, an NADH-dependent enzyme. The product is β -ketoacyl-CoA.

4. The enzyme thiolase catalyzes the cleavage of the b-ketoacyl-CoA; a molecule of CoA is required for the reaction. The products are acetyl-CoA and an acyl-CoA that is two carbons shorter than the original molecule that entered the f~oxidation cycle.

> The CoA is needed in this reaction to form the new thioester bond in the smaller acyl-CoA molecule. This smaller molecule then undergoes another round of the β-oxidation cycle.

When a fatty acid with an even number of carbon atoms undergoes successive rounds of the b-oxidation cycle, the product is acetyl-CoA. (Fatty acids with even numbers of carbon atoms are the ones normally found in nature, so acetyl-CoA is the usual product of fatty acid catabolism.) The number of molecules of acetyl-CoA produced is equal to half the number of carbon atoms in the original fatty acid.

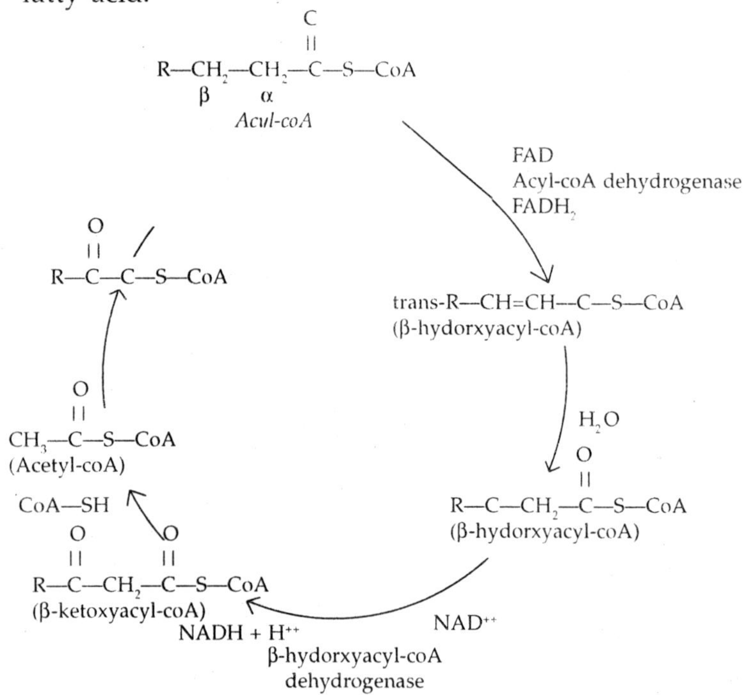

Fig.6. The b-oxidation cycle

The acetyl-CoA enters the citric acid cycle, with the rest

of the oxidation of fatty acids to carbon dioxide and water taking place through the citric acid cycle and electron transport. Recall that most of the enzymes of the citric acid cycle are located in the mitochondrial matrix, and we have just seen that the B-oxidation cycle takes place in the matrix as well. In addition to mitochondria, other sites of B-oxidation are known.

Peroxisomes and glyoxysomes, organelles that carry out oxidation reactions, are also sites of b-oxidation albeit to a far lesser extent than the mitochondria.

Oxidation of fatty acids

In carbohydrate metabolism the energy released by oxidation reactions is used to drive the production of ATP, with most of the ATP produced in aerobic processes. In the same aerobic processes—namely, the citric acid cycle and oxidative phosphorylation—the energy released by the oxidation of acetyl-CoA formed by b-oxidation of fatty acids can also be used to produce ATP. There are two sources of ATP to keep in mind when calculating the overall yield of ATP.

The first source is the reoxidation of the NADH and $FADH_2$ produced by the b b-oxidation of the fatty acid to acetyl-CoA. The second source is ATP production from the processing of the acetyl-CoA through the citric acid cycle and oxidative phosphorylation.

We shall use the oxidation of stearic acid, which contains 18 carbon atoms, as our example. Eight cycles of b-oxidation are required to convert 1 mole of stearic acid to 9 moles of acetyl-CoA; in the process 8 moles of FAD are reduced to $FADH_2$, and 8 moles of NAD^+ are reduced to NADH.

$$CH_3(CH_2)_{16}\overset{\overset{\displaystyle O}{||}}{C}—S—CoA + 8FAD + 8NAD^+ + 8H_2O + 8\ CoA\text{-}SH \rightarrow$$

$$9CH_3—\overset{\overset{\displaystyle O}{||}}{C}—S—CoA + 8FADH_2 + 8NADH + 8H^+$$

The 9 moles of acetyl-CoA produced from each mole of stearic acid enter the citric acid cycle. One mole of FADH2 and 3 moles of NADH are produced for each mole of acetyl-CoA that enters the citric acid cycle.

$$9\ CH_3\overset{\overset{\displaystyle O}{||}}{C}—S—CoA + 9FAD + 27NAD^+ + 9\ GDP + 9\ Pi + 27\ H_2O \rightarrow$$

At the same time, 1 mole of GDP is phosphorylated to produce GTP for each turn of the citric acid cycle. The $FADH_2$ and NADH produced by β-oxidation and by the citric acid cycle enter the electron transport chain, and ATP is produced by oxidative phosphorylation. In our example there are 17 moles of $FADH_2$, 8 from β-oxidation and 9 from the citric acid cycle; there are also 35 moles of NADH, 8 from β-oxidation and 27 from the citric acid cycle. Recall that 2.5 moles of ATP are produced for each mole of NADH that enters the electron transport chain, and 1.5 moles of ATP result from each mole of $FADH_2$.

Since $17 \times 1.5 = 25.5$, and 35 ´ 2.5 = 87.5, we can write the following equations:

$$17\ FADH_2 + 8.5\ O_2 + 25.5\ ADP + 25.5\ Pi \rightarrow 17\ FAD + 25.5\ ATP + 17\ H_2O$$

The overall yield of ATP from the oxidation of stearic acid can be obtained by adding the equations for b-oxidation, for the citric acid cycle, and for oxidative phosphorylation. In this calculation we take GDP as equivalent to ADP and GTP as equivalent to ATP, which means that the equivalent of nine ATP must be added to those produced in the reoxidation of $FADH_2$ and NADH. There are 9 ATP equivalent to the 9 GTP from the citric acid cycle, 25.5 ATP from the reoxidation of $FADH_2$, and 87.5 ATP from the reoxidation of NADH, for a grand total of 122 ATP.

The activation step in which stearyl-CoA was formed is not included in this calculation, and we must subtract the ATP that was required for that step. Even though only one ATP was required, two "high-energy" phosphate bonds are lost because of the production of AMP and PPi.

The pyrophosphate must be hydrolyzed to phosphate (P) before it can be recycled in metabolic intermediates. As a result we must subtract the equivalent of two ATP for the activation step. The net yield of ATP becomes 120 moles of ATP for each mole of stearic acid that is completely oxidized.

The yield of ATP from the oxidation of the lipid is still higher than that from the carbohydrate, even for the same number of carbon atoms. The reason is that a fatty acid is all hydrocarbon except for the carboxyl group; that is, it exists in a highly reduced state. A sugar is already partly oxidized because of the presence of its oxygen-containing groups.

Water is produced in the oxidation of fatty acids. We have already seen that water is also produced in the complete oxidation of carbohydrates. The production of metabolic water is a common feature of aerobic metabolism. This process can be a source of water for organisms that live in desert environments. Camels are a well-known example; the stored lipids in their humps are a source of both energy and water during long trips through the desert.

The kangaroo rat is an even more striking example of adaptation to an and environment. This animal has been observed to live indefinitely without having to drink water. It lives on a diet of seeds, which are rich in lipids but contain little water. The metabolic water that the kangaroo rat produces is adequate for all its water needs. This metabolic response to and conditions is usually accompanied by a reduced output of urine.

Additional reactions in oxidation of fattyacids

Fatty acids with odd numbers of carbon atoms are not as frequently encountered in nature as are the ones with even numbers of carbon atoms. Odd-numbered fatty acids also undergo the b-oxidation process. The last cycle of P-oxidation produces one molecule of propionyl-CoA.

An enzymatic pathway exists to convert propionyl-CoA to succinyl-CoA, which then enters the citric acid cycle. In this pathway, propionyl-CoA is first carboxylated to methyl malonyl-CoA in a reaction catalyzed by propionyl-CoA carboxylase, which then undergoes rearrangement to form succinyl-CoA; since propionyl-CoA is also a product of the catabolism of several amino acids, the conversion of propionyl-CoA to succinyl-CoA also plays a role in amino acid metabolism.

The conversion of unsaturated fatty acids to acetyl-CoA requires two reactions that are not encountered in the oxidation of saturated acids, a cis-trans isomerization and an epimerization. Successive rounds of β-oxidation of linoleic acid, which has two double bonds, provides an example of these reactions.

The process of β-oxidation gives rise to unsaturated fatty acids in which the double bond is in the trans arrangement, whereas the double bonds in most naturally occurring fatty acids are in the cis arrangement; in the case of linoleic acid there are two cis double bonds, between carbons 9 and 10 and between carbons 12 and 13.

Three rounds of β-oxidation produce a 12-carbon unsaturated fatty acid with the two cis double bonds between carbons 3 and 4 and between carbons 6 and 7. The hydratase of the α-oxidation cycle requires a trans double bond between carbon atoms 2 and 3 as a substrate. A cis-trans isomerase produces a trans double bond between carbons 2 and 3 from the cis double bond between carbons 3 and 4.

Two more cycles of β-oxidation give rise to an eight-carbon fatty acid with a is double bond between carbons 2 and 3. The position of the double bond makes the eight-carbon intermediate a possible substrate for the hydratase, but the stereochemistry of the product presents a problem in the next step of the cycle. The α-hydroxyl product has the D configuration, and the dehydrogenase that catalyzes the next stage of β-oxidation requires the L configuration. An epimerase catalyzes the required change of configuration, and the oxidation of the fatty acid can proceed. Unsaturated fatty acids make up a large enough portion of the fatty acids in storage fat to make the reactions of the cis-trans isomerase and the epimerase of some importance.

Formation of ketone bodies

Ketone bodies are produced when an excess of acetyl-CoA arises from β-oxidation. This condition occurs when not enough oxaloacetate is available to react with the large amounts of acetyl-CoA that could enter the citric acid cycle. Oxaloacetate in turn arises from glycolysis since it is formed from pyruvate in a reaction catalyzed by pyruvate carboxylase. A situation like this can come about when an organism has A high intake of lipids and a low intake of carbohydrates, but there are other possible causes as well, such as starvation and diabetes.

Starvation conditions cause an organism to break down fats for energy, leading to the production of large amounts of acetyl-CoA by β-oxidation. The amount of acetyl-CoA is excessive by comparison with the amount of oxaloacetate available to react with it. In the case of diabetics, the cause of the imbalance is not inadequate intake of carbohydrates but rather the inability to metabolize them. The reactions that result in "ketone bodies" start with the condensation of two molecules of acetyl-CoA to produce acetoacetyl-CoA.

Acetoacetate arises from acetoacetyl-CoA and can have two possible fates. A reduction reaction can produce β-hydroxybutyrate from acetoacetate. The other possible reaction is the spontaneous decarboxylation of acetoacetate to give acetone. The odour of acetone can frequently be detected on the breath of diabetics whose disease is not controlled by suitable treatment. The excess of acetoacetate, and consequently of acetone, is a pathological condition known as ketosis.

Since acetoacetate and β-hydroxybutyrate are acidic, their presence at high concentration overwhelms the buffering capacity of the blood. The consequent lowering of blood pH (ketoacidosis) is dealt with by the body by excreting H^+ into the urine, accompanied by excretion of Na^+, K^+ and water. Severe dehydration can result (excessive thirst is a classic symptom of diabetes); diabetic coma is another possible danger.

The principal site of synthesis of ketone bodies is liver mitochondria, but they are not used there. It is easy to transport ketone bodies in the bloodstream because, unlike fatty acids, they are water soluble and do not need to be bound to proteins such as serum albumin. Organs other than the liver use ketone bodies, particularly acetoacetate. Even though glucose is the usual fuel in most tissues and organs, acetoacetate can be and is used as a fuel.

In heart muscle and the renal cortex, acetoacetate is the preferred source of energy. Even in organs such as the brain, in which glucose is the preferred fuel, starvation conditions can lead to the use of acetoacetate for energy. In this situation acetoacetate is converted to two molecules of acetyl-CoA, which can then enter the citric acid cycle.

Ketosis and weight loss

Consistent overeating invariably results in weight gain, no matter what the source of the excess calories may be: carbohydrates, lipid, protein, or even alcohol. When you gain weight, your adipose tissue tends both to add more cells and to enlarge the ones already there. Severely obese people are unequivocally known to have added cells. Obese children, even ones who are not necessarily extremely overweight, also have more adipose cells. Obese children retain those added cells for life. During weight loss, the existing cells merely shrink, so there are lots of little, "starving" cells just waiting for a chance to grow again. The existence of these starving cells may explain why many people have a tendency to regain weight so easily.

Most of the time, the body cells depend on glucose from the blood for their energy. Overnight, when carbohydrate levels in the blood decrease, blood proteins are degraded in the liver to convert them to glucose; adipose tissue is also degraded to some extent. In the course of a diet, the body begins to break down muscle mass (protein) for conversion to glucose. After a few days to a week, nitrogen depletion becomes excessive. The body turns away from sources of glucose and turns to fats in adipose tissue. These fats are metabolized to fatty acids and glycerol, with the glycerol available for conversion to glucose, taking care of specialized tissues with a specific requirement for glucose.

The fatty acids are converted to short-chain oxy-acids usually called ketone bodies such as β-hydroxybutyrate and β-ketobutyrate, which are highly water soluble and are circulated easily in the blood. The ketone bodies are excellent and efficient nutrients, since they enter directly into the mitochondria for aerobic metabolism. The heart uses ketone bodies all the time.

Adipose tissue surrounds the heart and even permeates into it, providing a direct and efficient energy supply for this constantly working organ. The brain will adapt to the use of ketone bodies as nutrients after about a week of fasting. This is a surprising observation, since it had been believed that the brain could only use sugars. At this point, weight loss becomes very effective. Why do you lose weight so effectively?

Part of the answer depends on the fact that the ketone bodies are potentially dangerous, since they are relatively strong organic acids and they lower the blood pH, possibly causing acidosis. Recall that at low pH, hemoglobin is less able to bind oxygen. If you drink lots of fluids, your kidney will flush out the ketones in the urine to adjust body pH. You do lose weight very efficiently at this point, since you are literally putting excess calories "down the drain."

Prolonged ketosis/acidosis can be dangerous or even fatal. One reason is that the excess urination could produce a secondary mineral imbalance and this is dangerous, especially if one has a weak heart.

Medically supervised diets try to avoid a true ketosis state and people on such a limited calorie diet should be under constant physician care. You can tell you are in extreme ketosis by the metallic taste (acetone breath) in your mouth when you get up in the morning.

The existence of ketone bodies, the intermediate products of rapid metabolism of adipose tissue, has been known for a long time. The significance of these compounds both in normal metabolism and during dieting and starvation has not always been appreciated. Ketone bodies are important in weight loss and starvation; their presence can be readily monitored in the urine. In fact, the presence of ketone bodies in the urine of a person with diabetes mellitus is a clear indication that the disease is not under control.

We now recognize that ketone bodies represent a major and regular source of energy to the heart, probably the most consistently working organ in the body. The adipose tissue that surrounds the heart in essence "predigests" dietary carbohydrate to form ketone bodies.

A healthy heart never works anaerobically, and the ketone bodies offer the specific advantage of being absorbed directly into the mitochondria with no intermediate processing and no carrier molecules required. In other words, not only are ketone bodies a good source of energy, the are used very efficiently.

Anabolism of fatty acids

The anabolism of fatty acids is not simply a reversal of the reactions of β-oxidation. Anabolism and catabolism are not, in general, the exact reverse of each other; for instance, gluconeogenesis is not simply a reversal of the reactions of glycolysis. A first example of the differences between the degradation and the biosynthesis of fatty acids is that the anabolic reactions take place in the cytosol. We have just seen that the degradative reactions of β-oxidation take place in the mitochondrial matrix.

The first step in fatty acid biosynthesis is transport of acetyl-CoA to the cytosol. Acetyl-CoA can be formed either by β-oxidation of fatty acids or by decarboxylation of pyruvate. Most of these reactions take place in the mitochondria, requiring a transport mechanism to export acetyl-CoA to the cytosol for fatty acid biosynthesis. The transport mechanism is based on the fact that citrate can cross the mitochondrial membrane.

Acetyl-CoA condenses with oxaloacetate (which cannot cross the mitochondrial membrane) to form citrate (recall that this is the first reaction of the citric acid cycle). The citrate that is exported to the cytosol can undergo the reverse reaction, producing oxaloacetate and acetyl-CoA. Acetyl-CoA enters the pathway for fatty acid biosynthesis, whereas oxaloacetate undergoes a series of reactions in which there is a substitution of NADPH for NADH. This substitution exercises control over the pathway, since NADPH is required for fatty acid anabolism.

In the cytosol, acetyl-CoA is carboxylated, producing malonyl-CoA, a key intermediate in fatty acid biosynthesis. This reaction is catalyzed by the acetyl-CoA carboxylase complex, which consists of three enzymes and which requires Mn^{2+} and biotin for activity, in addition to ATP. We have already seen that enzymes catalyzing reactions that take place in several steps frequently consist of several separate protein molecules, and this enzyme follows that pattern.

In this case, acetyl-CoA carboxylase consists of the three proteins biotin carboxylase, the biotin carrier protein, and carboxyl transferase. Biotin carboxylase catalyzes the transfer of the carboxyl group to biotin.

The "activated CO_2 (the carboxyl group derived from the bicarbonate ion HCO_2) is covalently bound to biotin. Biotin (whether carboxylated or not) is bound to the biotin carrier protein by an amide linkage to the E-amino group of a lysine side chain. The amide linkage to the side chain that bonds biotin to the carrier protein is long enough and flexible enough to move the carboxylated biotin into position to transfer the carboxyl group to acetyl-CoA in the reaction catalyzed by carboxyl transferase, producing malonyl-CoA.

The biosynthesis of fatty acids involves the successive addition of two-carbon units to the growing chain. Two of the three carbon atoms of the malonyl group of malonyl-CoA are added to the growing fatty acid chain at each stage of the biosynthetic reaction. This reaction, like the formation of the malonyl-CoA itself, requires a multienzyme complex located in the cytosol and not attached to any membrane.

The complex, made up of the individual enzymes, is called fatty acid synthase. The usual product of fatty acid anabolism is palmitate, the 16-carbon saturated fatty acid. All 16 carbons come from the acetyl group of acetyl-CoA; we have already seen how malonyl-CoA, the immediate precursor, arises from acetyl-CoA. But first there is a priming step in which one molecule of acetyl-CoA is required for each molecule of palmitate produced.

In this priming step the acetyl group from acetyl-CoA is transferred to an acyl carrier protein (ACP), which is considered a part of the fatty acid synthase complex. The acetyl group is bound to the protein as a thioester. The group on the protein to which the acetyl group is bonded is the 4' -phosphopantetheine group, which in turn is bonded to a serine side chain.

The acetyl group is transferred in- turn from the ACP to another protein, to which it is bound by a thioester linkage to a cysteine-SH; the other protein is β-ketoacyl-S-ACP-synthase. The first of the successive additions of two of the three malonyl carbons to the fatty acid starts at this point. The malonyl group itself is transferred from a thioester linkage with CoA-SH to another thioester bond to the ACP.

The next step is a condensation reaction that produces acetoacetyl-S-ACP. In other words, the principal product of this reaction is an acetoacetyl group bound to the ACP by a thioester linkage. Two of the four carbons of acetoacetate come from the priming acetyl group, and the other two come from the malonyl group. The carbon atoms that arise from the malonyl group are the one directly bonded to the sulphur and the one in the -CH_2-group next to it.

The CH_3-CO-group comes from the priming acetyl group. The other carbon of the malonyl group is released as CO_2; the CO_2 that is lost is the original CO_2 that was used to carboxylate the acetyl-CoA to produce malonyl-CoA. The synthase is no longer involved in a thioester linkage.

Acetoacetyl-ACP is converted to butyryl-ACP by a series of reactions involving two reductions and a dehydration. In the first reduction the β-keto group is reduced to an alcohol, giving rise to D-β-hydroxybutyryl-ACP. In the process NADPH is oxidized to $NADP^+$ the enzyme that catalyzes this reaction is β– ketoacyl-ACP reductase. The dehydration step, catalyzed by β-hydroxyacyl-ACP dehydratase, produces crotonyl-ACP.

A second reduction reaction, catalyzed by β-enoyl-ACP reductase, produces butyryl-ACP. In this reaction NADPH is the coenzyme, as it was in the first reduction reaction in this series. In the second round of fatty acid biosynthesis, butyryl-ACP plays the same role as acetyl-ACP in the first round. The butyryl group is transferred to the synthase, and a malonyl group is transferred to the ACP. Once again there is a condensation reaction with malonyl-ACP.

In this second round the condensation produces a six-carbon β-ketoacyl-ACP. The two added carbon atoms come from the malonyl group, as they did in the first round. The reduction and dehydration reactions take place as before, giving rise to hexanoyl-ACP. The same series of reactions is repeated until palmitoyl-ACP is produced.

In mammalian systems the process stops at C_{16}, since the fatty acid synthase does not produce longer chains. Mammals produce longer-chain fatty acids by modification of the fatty acids formed by the synthase reaction. Fatty acid synthases from different types of organisms have markedly different characteristics.

In Escherichia coli the multienzyme system consists of an aggregate of separate enzymes, including a separate ACP. The ACP is of primary importance to the complex and is considered to occupy a central position in it. The phosphopantetheine group plays the role of a "swinging arm," much like that of biotin. This bacterial system has been extensively studied and has been considered a typical example of a fatty acid synthase.

The biosynthesis of fatty acids proceeds by the addition of two-carbon units to the hydrocarbon chain. The process takes place in the cytosol and is catalyzed in many organisms by a large multienzyme complex.

In mammals and in yeast, however, the fatty acid synthase consists of only two subunits (two different kinds of polypepticle chain), in contrast to the larger number in the bacterial system. Each of the subunits is a multifunctional enzyme that catalyzes reactions requiring several different proteins in the E. coli system.

The ACP and the enzyme for the condensation reaction appear to be the two subunits in the mammalian and yeast systems. The growing fatty acid chain swings back and forth between the two on the "swinging arm, and different parts of these two polypeptide chains catalyze the other stages of the addition of each two-carbon unit. Like the bacterial system, the eukaryotic system keeps all the components of the reaction in proximity to one another.

Several additional reactions are required for the elongation of fatty acid chains and the introduction of double bonds. When mammals produce fatty acids with longer chains than that of palmitate, the reaction does not involve cytosolic fatty acid synthase. There are two sites for the chain-lengthening reactions: the encloplasmic reticulum (ER) and the mitochondrion. In the chain-lengthening reactions in the mitochondrion, the intermediates are of the acyl-CoA type rather than the acyl-S-ACP type.

In other words, the chainlengthening reactions in the mitochondrion are the reverse of the catabolic reactions of fatty acids, with acetyl-CoA as the source of added carbon atoms; this is a difference between the main pathway of fatty acid biosynthesis and these modification reactions. In the ER the source of additional carbon atoms is malonyl-CoA. The modification reactions in the ER also differ from the biosynthesis of palmitate in that, like the mitochondrial reaction, there are no intermediates bound to ACP.

Reactions in which a double bond is introduced in fatty acids mainly take place on the ER. The insertion of the double bond is catalyzed by an oxidase that requires molecular oxygen (O_2) and NADH. Reactions linked to molecular oxygen are comparatively rare.

Mammals cannot introduce a double bond beyond

carbon atom 9 (counting from the carboxyl end) of the fatty acid chain.

As a result, linoleate with two double bonds, and linolenate with three double bonds, must be included in the diets of mammals. They are essential fatty acids because they are precursors of other lipids, including prostaglandins. Even though both the anabolism and the catabolism of fatty acids require successive reactions of two-carbon units, the two pathways are not the exact reversal of each other.

Anabolism of acylglycerols

Triacylglycerols, phosphoacylglycerols, and steroids, are derived from fatty acids and metabolites of fatty acids such as acetoacetyl-CoA. Free fatty acids do not occur in the cell to any great extent; they are normally found incorporated in triacylglycerols and phosphoacylglycerols. The biosynthesis of these two types of compounds takes place principally on the ER of liver cells or fat cells (adipocytes).

Triacylglycerols

The glycerol portion of lipids is derived from glycerol-3-phosphate, a compound available from glycolysis. Another source is glycerol released by degradation of acylglycerols. An acyl group of a fatty acid is transferred from an acyl-CoA. The products of this reaction are CoA-SH and a lysophosphatidate (a monoacylglycerol phosphate). The acyl group is shown as esterified at carbon atom 2 (C-2) in this series of equations, but it is equally likely that it is esterified at C-1.

A second acylation reaction takes place, catalyzed by the same enzyme, producing a phosphatidate (a diacylglyceryl phosphate). Phosphatidates occur in membranes and are precursors of other phospholipids. The phosphate group of the phosphatidate is removed by hydrolysis, producing a diacylglycerol. A third acyl group is added in a reaction in which the source of the acyl group is an acyl-CoA rather than the free fatty acid.

Phosphoacylglycerols

Phosphoacylglycerols (phosphoglycefides) are based on phosphatidates, with the phosphate group esterified to another alcohol, frequently a nitrogen-containing alcohol such as ethanolamine. The conversion of phosphatidates to other phospholipids frequently requires the presence of nucleoside triphosphates, particularly cytidine triphosphate (CTP).

The role of CTP depends on the type of organism, since the details of the biosynthetic pathway are not the same in mammals and bacteria. We shall use a comparison of the synthesis of phosphatidylethanolamine in mammals and in bacteria as a case study of the kinds of reactions commonly encountered in phosphoglyceride biosynthesis. In bacteria CTP reacts with phosphatidate to produce cytidine diphosphodiacylglycerol (a CDP diglyceride). The CDP diglyceride reacts with ethanolamine phosphate to form phosphatidylethanolamine.

In mammals the synthesis of phosphatidylethanolamine requires two preceding steps in which the component parts are processed. The first of these two steps is the removal by hydrolysis of the phosphate group of the phosphatidate, producing a diacylglycerol; the second step is the reaction of ethanolamine phosphate with CTP to produce pyrophosphate (PPj) and cytidine diphosphate ethanolamine (CDP-ethanolamine). The CDP-ethanolamine and diacylglycerol react to form phosphatidylethanolamine.

Sphingolipids

The structural basis of sphingolipids is not glycerol but sphingosine, a long chain amine. The precursors of sphingosine are palmitoyl-CoA and the amino acid serine, which react to produce dihydrosphingosine. The carboxyl group of the serine is lost as CO_2 in the course of this reaction. An oxidation reaction introduces a double bond, with sphingosine as the resulting compound. Reaction of the amino group of sphingosine with another acyl-CoA to form an amide CoA-SH bond results in an N-acylsphingosine, also called a ceramide.

Ceramides in turn are the parent compounds of sphingomyelins, cerebrosides, and gangliosides. Attachment of phosphorylcholine to the primary alcohol group of a ceramide produces a sphingomyelin, whereas attachment of sugars such as glucose at the same site produces cerebrosides. Gangliosides are formed from ceramides by attachment of oligosaccharides that contain a sialic acid residue, also at the primary alcohol group.

Anabolism of cholesterol

The ultimate precursor of all the carbon atoms in cholesterol and in the other steroids that are derived from cholesterol is the acetyl group of acetyl CoA. There are many steps in the biosynthesis of steroids. The condensation of three acetyl groups produces mevalonate, which contains six carbons. De carboxylation of mevalonate produces the five-carbon isoprene unit frequently encountered in the structure of lipids. The involvement of isopreneunits is a key point in the biosynthesis of steroids and of many other compounds that have the generic name terpenes.

Vitamins A, E, and K come from these reactions, which humans cannot carry out. That is why these compounds are vitamins that we require; vitamin D, the remaining lipid-COA-SH soluble vitamin, is derived from cholesterol.

Isoprene units are involved in the biosynthesis of ubiquinone (coenzyme Q) and of derivatives of proteins and tRNA with specific five-carbon units attached. Six isoprene units condense to form squalene, which contains 30 carbon atoms. Finally squalene is converted to cholesterol, which contains 27 carbon atoms; squalene can also be converted to other sterols.

$$\text{Acetate} \rightarrow \text{mevalonate} \rightarrow [\text{isoprene}] \rightarrow \text{squalene} \rightarrow \text{cholesterol}$$

12 of the carbon atoms of cholesterol arise from the carboxyl carbon of the acetyl group. The other 15 carbon atoms arise from the methyl carbon of the acetyl group; these are the carbon atoms labelled "m." We shall now look at the individual steps of the process in more detail.

Fig. 8. Biosynthesis of cholesterol.

The conversion of three acetyl groups of acetyl-CoA to mevalonate takes place in several steps. We already saw the first of these steps, the production of acetoacetyl-CoA from two molecules of acetyl-CoA, when we discussed the formation of ketone bodies and the anabolism of fatty acids. A third molecule of acetyl-CoA condenses with acetoacetyl-CoA to produce β-hydroxy-β-methylglutaryl-CoA (also called HMG-CoA and β-hydroxy-β-methylglutaryl-CoA).

This reaction is catalyzed by the enzyme hydroxymethylglutaryl-CoA synthase; one molecule of CoA-SH is released in the process. In the next reaction the production of mevalonate from hydroxy-methylglutaryl-CoA is catalyzed by the enzyme hydroxymethylglutaryl-CoA reductase.

A carboxyl group, the one esterified to CoA-SH, is reduced to a hydroxyl group, and the CoA-SH is released. This step is inhibited by high levels of cholesterol. It is also a target for drugs to lower cholesterol levels in the body. The drug lovastatin is an inhibitor of hydroxymethyl-CoA reductase; it is widely prescribed to lower blood cholesterol levels.

Mevalonate is then converted to an isoprenoid unit by a combination of phosphorylation, decarboxylation, and dephosphorylation reactions. Three successive phosphorylation reactions, each of which is catalyzed by an enzyme that requires ATP, give rise to 3-phospho-5-pyrophosphomevalonate. This last compound is unstable, and it undergoes a decarboxylation and a dephosphorylation reaction to produce isopentenyl pyrophosphate, a five-carbon isoprenoid derivative.

Isopentenyl pyrophosphate and dimethylallyl pyrophosphate, another isoprenoid derivative, can be interconverted in a rearrangement reaction catalysed by the enzyme isopentenyl pyrophosphate isomerase. Condensation of isoprenoid units then leads to the production of squalene and ultimately of cholesterol. Both of the isoprenoid derivatives we have met so far are required; isopentenyl pyrophosphate and dimethylallyl pyrophosphate condense with each other in a reaction catalyzed by dimethylallyl transferase (also called prenyl transferase) to produce geranyl pyrophosphate, a ten-carbon compound.

Another condensation reaction takes place, this time between geranyl pyrophosphate and isopentenyl pyrophosphate. This reaction is again catalyzed by dimethylallyl transferase; farnesyl pyrophosphate, a 15-carbon compound, is produced. Two molecules of farnesyl pyrophosphate condense to form squalene, a 30-

carbon compound. The reaction is catalyzed by squalene synthase, and NADPH is required for the reaction.

Conversion of squalene to cholesterol are far from simple. Squalene is converted to squalene epoxide in a reaction that requires both NADPH and molecular oxygen (O_2). This reaction is catalyzed by squalene monooxygenase. Squalene epoxide then undergoes a complex cyclization reaction to form lanosterol. This remarkable reaction is catalyzed by squalene epoxide cyclase. The mechanism of the reaction is a concerted reaction, that is, one in which each part is essential for any other part to take place.

No portion of a concerted reaction can be left out or changed, because it all takes place simultaneously rather than a sequence of steps. The conversion of lanosterol to cholesterol is a complex process. It is known that 19 steps are required to remove three methyl groups and to move a double bond, but we shall not discuss the details of the process.

Once cholesterol is formed, it can be converted to other steroids of widely varying physiological function. The smooth ER is an important site for both the synthesis of cholesterol and its conversion to other steroids. Most of the cholesterol formed in the liver, which is the principal site of cholesterol synthesis in mammals, is converted to bile acids such as cholate and glycocholate. These compounds aid in the digestion of lipid droplets by emulsifying them and rendering them more accessible to enzyme attack. Cholesterol is the precursor of important steroid hormones, in addition to the bile acids.

Like all hormones, whatever their chemical nature, steroid hormones serve as signals from outside a cell that regulate metabolic processes within a cell. Steroids are best known as sex hormones (they are a component of birth control pills), but they play other roles as well.

Pregnenolone is formed from cholesterol, and progesterone is formed from pregnenolone. Progesterone is a sex hormone and is a precursor for other sex hormones such as testosterone and estradiol. (an estrogen). Other types of steroid hormones also arise from progesterone.

Cortisone is an example of glucocorticoids, a group of hormones that play a role in carbohydrate metabolism, as the name implies, as well as in the metabolism of proteins and fatty acids. Mineralocorticoids constitute another class of hormones that are involved in the metabolism of electrolytes, including metal ions, and water. Aldosterone is an example of a mineralocorticoid. In cells in which cholesterol is converted to steroid hormones, an enlarged smooth ER is frequently observed, providing a site for the process to take place.

The biosynthesis of cholesterol proceeds by the condensation of five-carbon isoprenoid units to form the key intermediate squalene isoprenoid units in turn are derived from the reaction of three acetyl-CoA units. Once cholesterol is formed, it serves as the precursor for other steroids.

Atherosclerosis is a condition in which arteries are blocked to a greater or lesser extent by the deposition of cholesterol plaques, leading to heart attacks. The process by which the clogging of arteries occurs is a complex one. Both diet and genetics are instrumental in the development of atherosclerosis.

A diet high in cholesterol and fats, particularly saturated fats, will lead to a high level of cholesterol in the bloodstream. The body also makes its own cholesterol because this steroid is a necessary component of cell membranes. It is possible for more cholesterol to come from endogenous sources (synthesized within the body) than from the diet. Cholesterol must be packaged for transport in the bloodstream; several classes of lipoproteins are involved in the transport of lipids in blood.

These lipoprotein aggregates are usually classified by their densities. Besides chylomicrons, they include very low-density lipoproteins (VLDL), intermediate-density lipoproteins (IDL), low-density lipoproteins (LDL), and high-density lipoproteins (HDL). LDL and HDL will play the major role in our discussion of heart disease. The protein portions of these aggregates can vary widely. The major lipids are generally cholesterol and its esters, in which the hydroxyl group is esterified to a fatty acid; triacylglycerols -are also found in these aggregates.

Chylomicrons are involved in the transport of dietary lipids, whereas the other lipoproteins primarily deal with endogenous lipids.

The interior of LDL particle consists of many molecules of cholesteryl esters (the hydroxyl group of the cholesterol is esterified to an unsaturated fatty acid such as linoleate); on the surface, a protein (apoprotein B-100), phospholipids, and unesterified cholesterol are in contact with the aqueous medium of the plasma.

The protein portions of LDL particles bind to receptor sites on the surface of a typical cell. LDL particles are degraded in the cell. LDL particles are taken into the cell by the highly regulated process of endocytosis, in which a portion of the cell membrane containing the LDL particle and its receptor enters the cell. The receptor is returned to the cell surface, while the LDL particles are degraded in the lysosomes. The protein portion of LDL is hydrolyzed to the component amino acids, whereas the cholesterol esters are hydrolyzed to cholesterol and fatty acids.

Cholesterol not needed for membrane synthesis can be stored as oleate or palmitoleate esters in which the fatty acid is esterified tŏ the hydroxyl group of cholesterol. The production of these esters is catalyzed by acyl-CoA cholesterol acyltransterase (ACAT), and the presence of free cholesterol increases the enzymatic activity of ACAT. In addition, cholesterol inhibits both the synthesis and the activity of the enzyme hydroxymethylglutaryl-CoA reductase (HMG-CoA reductase). This enzyme catalyzes the production of mevalonate, the reaction that is the committed step in cholesterol biosynthesis.

Dietary cholesterol suppresses the synthesis of cholesterol by the body, especially in tissues other than the liver. The crucial role of LDL receptors in maintaining the level of cholesterol in the bloodstream is especially clear in the case of familial hypercholes terolemia, which results from a defect in the gene that codes for active receptors.

An individual who has one gene that codes for active receptor and one defective gene is heterozygous for this trait (a heterozygote). Heterozygotes have blood cholesterol levels that are above average and therefore are at higher risk for heart disease than the general population. An individual with two defective genes and thus no active LDL receptor is homozygous (a homozygote).

Homozygotes have very high blood cholesterol levels from birth, and there are recorded cases of heart attacks in two-year-olds with this condition. Patients who are homozygous for familial hypercholesterolemia usually die before age 20. Another genetic abnormality involved in hypercholesterolernia is the one that gives rise to a faulty protein component of lipoprotein E, involved in the uptake of lipids by the cell. The unfortunate result is the same.

The high-density lipoprotein (HDL), which transports cholesterol from the liver to the rest of the body, HDL transports it back to the liver for degradation to bile acids. It is desirable to have low levels of cholesterol and LDL in the bloodstream, but it is also desirable to have as high a proportion of total cholesterol as possible in the form of HDL. It is well known that high levels of LDL and low levels of HDL are correlated with the development of heart disease.

Chapter 8

Biochemistry of Photosynthesis

The energy transduction reaction taking place in plants and ome photosynthetic bacteria is known as photosynthesis. It is central to life since it is the ultimate source of food for almost all organisms. The scale of photosynthesis is vast: on the Earth some 2×10^{11} tonnes of carbon are fixed each year (> 6000 $tonness^{-1}$). An approximate equation for the overall process of photosynthesis can be written as

$$CO_2 + 2H_2O \rightarrow (CH_2O) + H_2O + O_2 \; \Delta G = 4.8 \times 10^5 \text{ J mol}^{-1}$$

Research into photosynthesis began more than 200 years ago, when Priestley discovered that photosynthesis was the reverse of respiration. The discoveries of the next century indicated that water, carbon dioxide, light and the green parts of plants were all necessary for photosynthesis.

The green pigment chlorophyll was isolated early in the nineteenth century, and later the hypothesis that photosynthesis led to the storage of chemical energy was proposed. There followed careful work on the stoichiometry of gas exchange, and the finding that starch was the product of photosynthesis. Therefore, towards the latter half of the nineteenth century, photosynthesis could be described by the approximate Equation given above. Then the elegant experiments of Englemann using oxygen –requiring motile bacteria and the green alga Spirogyra showed that the chloroplast was the seat of photosynthesis in the cell. All higher plants and algae, together with certain species of bacteria, are capable of storing light energy via photosynthesis.

The most primitive photosynthesizing cells are the photosynthetic bacteria, and their photosynthesis differs from that of other organisms because they are not able to evolve oxygen and because they contain the pigment bacteriochlorophyll, rather than chlorophyll as in green plants.

They are, however, widely used in photosynthesis research because the cells are amenable to a range of biochemical and molecular biological treatments. Mainly for this reason, our knowledge of the photosynthetic processes in bacteria is more advanced than that of higher plants. The link between bacterial photosynthesis and green–plant photosynthesis is provided by the blue–green algae (cyanobacteria).

These organisms have the characteristics of prokaryotic cells, but have the ability to evolve oxygen; their pigments are also more like those of green plants, since they contain chlorophyll.

Oxidation and reduction reactions

The approximate Equation for photosynthesis does not indicate whether water or carbon dioxide is the source of the oxygen evolved in green plant photosynthesis. Three lines of evidence suggest that water is oxidized to give molecular oxygen.

van Niel's hypothesis

van Niel studied the comparative biochemistry of photosynthesis in oxygen–evolving organisms and in the non–oxygen-evolving photosynthetic bacteria. The latter can be divided into three classes depending on their requirements for the assimilation of carbon dioxide.

Two of these classes (the green– and the purple–sulfur bacteria) can use hydrogen sulfide as a substrate for growth, and produce sulfur as a by product:

$$2H_2S + CO_2 \rightarrow CH_2O + H_2O + 2S$$

The third class of bacteria (the purple non–sulfur bacteria) use organic substances like succinate as a substrate, giving fumarate as a by–product:

$$2COOHCH_2CH_2COOH + CO_2 \rightarrow CH_2O$$
$$+ H_2O + 2COOHCHCHCOOH$$

Taking both the above equations for green plants e can write

$$2H_2O + CO_2 \rightarrow CH_2O + H_2O + O_2$$

van Niel generalized the process of photosynthesis, and wrote a general equation:

$$2H_2A + CO_2 \rightarrow CH_2O + H_2O + 2A$$

Here H_2A is an oxidizable substrate. van Niel's hypothesis has two important consequences: firstly, it gives a logical relationship between apparently diverse photosynthetic processes; secondly, it implies that water is the source of photosynthetic oxygen and that carbon dioxide is reduced to carbohydrate.

The Hill reaction

A much more direct line of evidence came from work by Hill and Scarisbrick on isolated chloroplasts, which showed that the illuminated chloroplasts would evolve oxygen only in the presence of a reducible substance (an electron acceptor or 'Hill oxidant') such as potassium ferricyanide.

The evolution of oxygen and the reduction of the electron acceptor have identical stoichiometries, and occur in the absence of fixation of carbon dioxide. The results of this experiment indicate that the evolution of oxygen involves the oxidation of water coupled to the reduction of another substance.

In the leaf, this latter compound is carbon dioxide, but the isolation of chloroplasts frequently leads to the loss of the enzymes of the carbon dioxide fixation cycle and consequently oxygen cannot be evolved without an exogenous electron acceptor. It is only comparatively recently that techniques have been developed to isolate chloroplasts able to fix carbon dioxide at rates comparable to those found in leaves.

^{18}O studies

An obvious experiment, in the light of Hill's findings, was to use the stable, heavy isotope of oxygen (^{18}O) to discriminate between water and carbon dioxide as the source of photosynthetic oxygen. When ^{18}O became available, such an experiment was performed by Ruben, Randall, Kamen and Hyde, and the result suggested that water was the source of evolved oxygen.

Light and dark reactions

During photosynthesis water is oxidized and carbon dioxide is reduced, but they give no indication of where in the overall process light energy intervenes to drive the reaction. However, it is—possible to show that photosynthesis consists of a combination of light–requiring reactions (the 'light reactions') and non–light –requiring reactions (the 'dark reactions'). Blackman interpreted the hyperbolic light-intensity dependence of photosynthesis as evidence for a temperature–independent photochemical reaction coupled to a 'dark' reaction which is dependent both on temperature and on the level of carbon dioxide.

Later, in the classic experiments of Emerson and Arnold here the rate of photo synthesis was measured under flashing light, it was found that the yield of photosynthesis per flash depended not only on flash intensity, but also on the time interval between flashes. The time interval between flashes which gave the maximum yield varied with temperature, but was typically 100ms, suggesting the existence of a relatively slow dark reaction which took about 100ms to be complete.

The functional relationship between the 'light' and 'dark' reactions can be established by examining the requirements of the 'dark' reactions.

The 'dark' reactions comprise a complex cycle of enzyme –mediated reactions (the Calvin cycle) which catalyzes the reduction of carbon dioxide to sugar. As well as carbon dioxide the cycle requires reducing power in the form of reduced nicotinamide adenine dinucleotide phosphate (NADP) and chemical energy in the form of adenosine triphosphate (ATP). The reduced NADP (NADPH) and ATP are produced by the 'light' reactions.

It is thus possible to divide a description of photosynthesis into those reactions associated with the Calvin cycle and the fixation of carbon dioxide, and those reactions (capture of light by pigments, electron transport, photophosphorylation) which are directly driven by light.

Photosynthetic apparatus

The chloroplast is bounded by an outer double membrane or envelope, and contains a complex network of thylakoid membranes surrounded by a matrix called the stroma. The division of photosynthesis into 'light' and 'dark' reactions is reflected in this structure; the light–driven reactions all occur in or on the thylakoid membranes while the complex cycle of enzyme–mediated 'dark' reactions takes place in the stroma.

Pigments

The first step in the 'light' reactions of photosynthesis is the absorption of light by pigment molecules. Pigments are molecules which absorb visible, light strongly, and hence appear colored; photosynthetic pigments have extensive conjugated double bonds in the light–absorbing part of the molecule.

All oxygen–evolving organisms contain chlorophyll–a (chl–a). The molecule has two distinct parts: the 'head' group consists of a cyclic tetrapyrrole with a Mg ion complexed in the center, and has extensive conjugated double bonds. The 'tail' of the molecule is a phytol side–chain which has no role in the absorption of light.

Higher plants and green algae also contain chlorophyll –*b*, in which the $-CH_3$ group on ring II of Chl–*a* is replaced by a –CHO group. Two other forms of chlorophyll (Chl–c and Chl–d) are found in diatoms and some algae. Blue–green algae lack Chl–b.

Photosynthetic bacteria contain pigments similar to chlorophyll, called bacteriochlorophylls (BChl). BChl has two fewer double bonds than Chl, and has an absorption spectrum in an acetone– methanol mixture which shows a major absorption band in the near–infra–red at 770 nm.

The major type is BChl–a, but other types exist. Photosynthetic organisms do not contain chlorophyll alone. All contain one or more carotenoids, which are linear molecules containing a number of conjugated double bonds.

There is a wide variety of carotenoids, and they can be split into the purely hydrocarbon carotenes and the oxygenated carotenols. Their types and occurrence are listed in Table 1.

Table 1
Types and occurrence of photosynthetic pigments

Pigment	*Occurrence*
Chlorophyll *a*	All oxygen-evolving organisms
Chlorophyll *b*	Higher plants and green algae
Chlorophyll *c*	Diatoms and brown algae
Chlorophyll *d*	Red algae
α–carotene	Higher plants, most algae
β–carotene	Most plants, some algae
Luteol	Higher plants, green and red algae
Violaxanthol	Higher plants
Fucoxanthol	Brown algae, diatoms
Phycoerythrin	Red algae, some blue-green algae
Phycocyanin	blue-green algae, some red algae
Allophycocyanin	Blue-green algae, red algae
Bacteriochlorophyll *a*	Purple and green bacteria
Bacteriochlorophyll *b*	Some purple bacteria

The blue–green and red algae have distinctive colors which are due to the presence of bile pigments in their cells. Bile pigments have a linear tetrapyrrole as the chromophore, and in some cases a water–soluble pigment–protein complex can be extracted.

The phycoerythrins absorb in the green and yellow regions of the spectrum and hence appear red; the phycocyanins appear blue. The unusual absorption characteristics of these pigments have been useful in determining the functional role of the photosynthetic pigments; the unicellular red alga *Porphyridium cruentian* has been a particularly favored organism.

Functional organization

With such a variety of types of pigments it is clearly important to know how each one plays its part in the capture of light energy. They are subject to a high degree of organization, and this contributes in large measure to the efficiency of the early steps of photosynthesis.

Light harvesting

The experimental technique that led to the concept of light energy being transferred from one pigment to the next is called sensitized fluorescence. Fluorescence itself is an important property of a number of pigment molecules. If a pigment is dissolved in an organic solvent and then illuminated, the absorbed light energy has a limited number of routes via which it may be dissipated. One of these routes involves the re–emission of the light, normally at a wavelength somewhat longer than that of the exciting light, and the emitted light is called fluorescence.

The fluorescence emission spectrum is a characteristic feature of a pigment. Experiments by Dutton, Manning and Duggar using a diatom showed that light absorbed by the carotenoid fucoxanthin excited fluorescence from Chl–a. Similarly, in the early Duysens investigated the fluorescence properties of pigments in the unicellular red alga *Porphyridium* which, as Table 1 indicates, contains bile pigments.

He found that if the algae were illuminated with monochromatic radiation absorbed mainly by phycoerythrin then the emitted fluorescence had a fluorescence spectrum characteristic of Chl–*a*. Since Chl–*a* itself was not directly excited, this suggests that it must have been excited indirectly by excitation energy transfer from phycoerythrin. Working in this way with each pigment in turn, Duysens was able to draw up an excitation energy transfer map where light absorbed by the bile pigments is transferred to Chl–*a* through a kind of molecular bucket–brigade, and likewise Chl–*b* feeds light energy directly to Chl–*a*.

Moreover, the efficiencies of energy transfer to Chl – *a* were established they are generally rather high, but sometimes the transfer from carotenoids has an efficiency of only 20–50%. Because of the time needed for excitation energy to transfer from one pigment to the next there will be a short, but finite, lag–time between excitation of B–phycoerythrin by light and the emission of fluorescence by pigments which are 'downstream' and accept the transferred excitation.

Experiments designed to measure such lag–times were first performed with *Porphyridiurn* by Brody and Rabinowitch, but the most striking evidence has come with the advent of mode–locked lasers, capable of producing a train of intense pulses of light, each pulse having a duration of a few picoseconds (one picosecond = 10^{12} S; light travels approximately 0.3 mm in 1 ps).

Using a mode–locked laser and a rapidly–responding photodetector, Porter and his colleagues excited one of the bile pigments in Porphyridium (B–phycoerythrin) with a pulse of wavelength 530 nm and of 6 ps duration, and monitored the fluorescence emission from B–phycoerythrin itself at 576 nm, then from R–phycocyanin (640 nm), allophycocyanin (661 nm) and Chl –*a* (685 nm).

The result is a direct confirmation of Duysens' work. Intermolecular excitation energy transfer is an important part of the function of the photosynthetic pigments, enabling wavelengths of light which Chl cannot absorb to contribute to photosynthesis.

Prevention of the photodynamic effect

In organic solution chlorophyll is a labile pigment. If a bright green solution of chlorophyll in acetone is left in sunlight it is bleached quickly and irreversibly to a pate straw color. This reaction requires light and oxygen, both of which are available in the leaf, and yet chlorophyll in vivo is not decolorized. Why might this be'?

The precise mechanism of the bleaching of chlorophyll (the 'photodynamic effect') is not altogether clear, but it appears that the over-excitation of chlorophyll molecules by light results in the formation of so–called Chl triplet states. A triplet state is an excited state in which the spin of the excited electron has been reversed: the state is stable for several microseconds and in a reaction with molecular oxygen can form either singlet oxygen ($\Delta_g O_2$) or superoxide (O_2^-).

Both of these forms of oxygen are very reactive, and lead to irreversible damage not only to photosynthetic pigments but also to nucleic acids and other compounds. In the leaf the formation of triplet Chl appears to be followed very rapidly by the quenching of the chlorophyll triplet due to transfer of excitation energy to neighboring carotenoid molecules.

This transfer leads to the formation of a harmless carotenoid triplet, whose energy is dissipated to heat.

B-phycoerythrin

↓ 70–80%

R—phycocyanin

allophycocyanin 80-90%

↓

Chl-*b* → Chl-*a*

Fig. 1. Pathway of excitation transfer from the accessory pigments to Chl-a in the red alga Porphyridium cruentum.

Reactions of this type have been detected by monitoring the rapid transient absorbance changes due to carotenoid triplets in chloroplasts. The protective role of carotenoids is of vital importance: it may be the primary function of the pigments, with the light harvesting role being secondary.

In the absence of carotenoids a photosynthetic organism is at a distinct disadvantage: a mutant strain of the purple non–sulfur bacterium *Rhodopseudomonas sphaeroides* which lacks colored carotenoids must be grown strictly anaerobically since it has no mechanism to deal with bacteriochlorophyll triplet states.

Photosynthetic unit

In the early part of this century there was much discussion about the part played by chlorophyll in photosynthesis. A major step forward was made by Emerson and Arnold in the same series of experiments that led to the idea of 'light' and 'dark' reactions.

Emerson and Arnold used neon tubes with a flash duration of about 10 μs to illuminate algal suspensions and measured manometrically the amount of oxygen evolved by a long sequence of flashes. They calculated the oxygen yield per flash and found that it was related to the energy of the flash up to a point, but that the oxygen yield saturated at relatively low flash energy.

Taking the point at which saturation has just been reached, Emerson and Arnold analyzed the algae for their chlorophyll content, and calculated the number of chlorophyll molecules associated with the evolution of one molecule of oxygen. They found the number was about 2500, and that this number was fairly constant over a range of measurements. Another ratio must be introduced here: the number of quanta that are required to evolve one molecule of oxygen. After fierce controversy in the generally agreed that at least 8 quanta are required.

Thus a fundamental association of 2500/8 ≅ 300 chlorophyll molecules is concerned with the processing of each quantum of light. The group of about 300 chlorophyll molecules is termed a 'photosynthetic unit'; its significance is that of the 300 chlorophyll molecules, only one 'special' Chl is concerned with the processing of light energy, while the remainder act to absorb and channel light energy to the special chlorophyll.

In this way, the effective capture cross–section of the 'special' chlorophyll is increased some 300–fold. The concept of the photosynthetic unit received further support from some calculations by Gaffron and Wohl, who estimated the time taken for oxygen evolution to begin in dark–adapted cells exposed to weak light. They calculated a long lag–time, but experimentally found a much shorter one, and suggested that chlorophyll molecules were able in some way to pool their resources by exchanging excitation energy among themselves.

Mechanism of intermolecular energy transfer

The photosynthetic unit is a very important idea because it describes the high level of organization among photosynthetic pigments; however it is still unclear whether the photosynthetic unit is a statistical or a physical entity.

Nevertheless, the occurrence of excitation energy transfer between different molecules, and between like molecules, is beyond doubt.

The first physical description of such excitation transfers was given by Förster who suggested that the precise mechanism would depend on the distance between donor and acceptor molecules: if R is the intermolecular distance then for large values of R (≥ 1.5 nm), the efficiency of energy transfer would depend on R^{-6}, but for small values of R (≥ 1.5 mn) the donor and acceptor molecules can 'see' each other, and the efficiency of energy transfer would depend on R^{-3}.

As well as depending on intermolecular distance, the efficiency of energy transfer is also a function (a) of the degree to which the fluorescence spectrum of the donor molecule overlaps with the absorption spectrum of the acceptor molecule (the 'overlap integral') and (b) of intermolecular orientation, so that energy transfer is more likely when the dipole moments of donor and acceptor molecules are aligned. This is like rotating a transistor radio to get a strong signal from the transmitter.

Förster's theories have been reviewed and extended by Knox, and some experimental studies have been reported. In physical systems of pigment mixtures, Förster's R^{-6} transfer mechanism appears to describe the intermolecular energy transfer well, but in biological systems the picture at present is not as clear.

Absorption spectra of the photosynthetic apparatus

In higher plants there appear to be only two chemically distinct forms of Chi, Chl–*a* and Chl–*b*. In the red region of the spectrum the only pigments that absorb are the chlorophylls, so that it might be expected that the absorption spectrum of the photosynthetic apparatus would be relatively simple. But in practice this is not the case.

The absorption spectrum of a sample of isolated chloroplasts, and this clearly differs from the absorption spectrum of Chi in organic solvent. The difference can be understood by realizing that in the thylakoid membrane Chi does not exist in free solution, but as pigment–protein complexes.

The physical environment of a pigment may affect its absorption spectrum, and the absorption spectrum of Chi–*a* is shifted to the red to a greater or lesser extent by interaction with protein, giving rise to a number of Chi – *a* species (really Chi – *a* – protein complexes) with absorption maxima in the wavelength range 650 to 720 nm.

An attempt at identification of the pigment–protein complexes can be made from the absorption spectrum by assuming that each complex has a single Gaussian (bell–shaped) absorption profile, and that the observed spectrum is the sum of a number of such profiles. Computer–assisted curve–fitting procedures can then be used to determine the number, wavelength maximum and width of each component.

The task is made easier if the sample is cooled to 77 K, since the components have reduced half–width; moreover, a fourth derivative spectrum obtained by differentiating the absorption spectrum four time with respect to wavelength can help determine the number of components. In a similar way the number of fluorescent Chi–*a*–protein components can be identified by careful study of the low–temperature fluorescence emission spectrum of chloroplasts.

Photosystems and electron transport

The action spectrum of photosynthesis is not identical to the absorption spectrum of chlorophyll further supports the idea that other pigments absorb light which is used in driving photosynthesis.

Firstly, at the short-wavelength end of the spectrum (i.e., about 420-500nm) some wavelengths of light are relatively inefficient at exciting photosynthesis; this corresponds to the rather poor efficiency of excitation energy transfer between carotenoid and chlorophyll.

Secondly, at the long–wavelength end of the spectrum, light between 680 and 720nm is also inefficient at exciting photosynthesis, although the absorption due to chlorophyll in this spectral region is still strong. This deficiency in the ability of red light to excite photosynthesis was termed the 'red drop', and might be interpreted as evidence for the existence of a long–wavelength–absorbing pool of Chl which, although it absorbs light, does not participate in photosynthesis. This interpretation is, however, inconsistent with the results of further experiments conducted by Emerson and his colleagues.

Emerson enhancement

In continuing and extending studies on the action spectra of photosynthesis, Emerson, Chalmers and Cederstrand found that the low efficiency of far–red light in exciting photosynthesis could be increased by simultaneously illuminating the sample with a second beam of light of shorter wavelength, say, at 650 nm. Not only was the low efficiency of the far– red light made up to the expected level, but the yield of photosynthesis in the combined beams was greater than the sum of the yields of the beams applied separately.

This effect is known as Emerson enhancement, and can be given a numerical value, *E*, by the equation

$$E = \frac{\Delta O_2 \text{ (combined)} - \Delta O_2 \text{ (short - wavelength alone)}}{\Delta O_2 \text{ (long - wavelength alone)}}$$

Here ΔO_2 represents the rate of oxygen evolution. Emerson enhancement is as much a photobiological phenomenon as photosynthesis itself, so an Emerson enhancement is as much a photobiological phenomenon as photosynthesis itself, so an action spectrum was necessary to ascertain which pigments are involved.

The results were rather complicated, but in red algae (which lack Chl-*b*) it appeared that the action spectrum for enhancement was the same as the absorption spectrum of the accessory pigments. Thus Emerson proposed that there were two light–driven processes in photosynthesis, one sensitized by Chl–*a* and the other sensitized by the accessory pigments.

It proved impossible to generalize this proposal: close examination of the action spectra of enhancement in green algae revealed that one of the species of chlorophyll–*a* itself was involved in sensitizing enhancement.

This observation changed the picture: the idea of two light–driven reactions sensitized by different sets of pigments remained, but the sets of pigments were not divided as simply as originally envisioned by Emerson.

The two light–driven reactions were named by Duysens, Amesz and Kamp as Photosystem II (PSII) and Photosystem I (PSI). PSII is sensitized by several types of Chl–*a*–protein complexes and by Chl–b which is also present in a pigment–protein complex.

In red algae, light absorbed by the bile pigments is preferentially transferred to PSII. In all oxygen–evolving organisms the pigments sensitizing PSII absorb only up to about 690 nm. PSI is also sensitized by Chl–a–protein complexes, but in this case the maximum wavelength of absorption is about 720nm. PSI is not strongly sensitized by Chl–*b*. If the two reactions are consecutive then the overall rate of photosynthesis will be governed by the rate–limiting step.

The explanation for the 'red drop' now becomes clear, since at exciting wavelengths above about 690 nm PSII ceases to function, and hence the overall rate of photosynthesis decreases even though PSI still absorbs light and tries to work. The 'red drop' is eliminated when the second beam of light excites PSII and speeds up the rate –limiting step; enhancement occurs because some of this second beam excites PSI as well.

Evidence for two photosystems

The concept of two photosystems each sensitized by a different set of pigment molecules helped to resolve a number of outstanding problems in the literature at that time.

For example, several workers had studied the action spectrum of chlorophyll fluorescence in vivo, and found that in red algae light absorbed by the phycobilins was more efficient in exciting chlorophyll fluorescence than light absorbed by chlorophyll itself. The data were now interpreted as suggesting that there were two 'pools' of Chi–*a*, one which was functionally associated with the phycobilins, and which had a high fluorescence yield, and the other which was not excited by the phycobilins, and which had a low fluorescence yield.

In retrospect, it is seen that the Chi with a high fluorescence yield is associated with PSII, and that with the low fluorescence yield with PSI. Another curious observation was called 'chromatic transients'. In experiments designed to adjust the fluence rates of two wavelengths of light to produce an equal rate of photosynthesis in each, Blincks noted a transient increase in the rate of photosynthesis when one beam was switched off and the other switched on, suggesting that enhancement as described above was occurring, but that the two wavelengths of light need not be applied simultaneously to the sample.

Myers and French noted that enhancement could be elicited by flashes of different wavelengths separated by a time interval of several seconds. Taken together, these observations suggest that the photosystem excited by the one wavelength has the ability to store some intermediate which is required by the second photosystem for it to function fully. The intermediates are associated with oxidized or reduced components of an electron transport chain which connects the two photosystems.

Z–scheme

Evidence for two light–driven photoprocesses was now clear, but the connexion between them was as yet undiscovered. In principle two methods of connexion were possible, either in series with the products of one reaction becoming the reactants of the second, or in parallel, where the two reactions proceed side by side.

Cytochromes are complexes of protein and heme: the prosthetic group of the molecule is the heme, and this consists of a cyclic tetrapyrrole with an iron atom complexed in the center. Cytochromes were well known to undergo oxidation and reduction changes which were associated with changes in the cytochrome absorption spectrum.

Hill had found both a *c*–type cytochrome and a b-type cytochrome and suggested that the cytochromes might act as oxidizable and reducible intermediates connecting the two photoreactions in series via electron transport.

Absorption spectrophotometry of cytochromes

The proposal of Hill and Bendall received support from the work of Duysens et al. who used a sensitive absorption – spectrophotometric technique to determine the redox state of the c–type cytochrome (cyt–f) in the red alga *Porphyridium*.

The algae were illuminated with light at 680 run, which is absorbed mainly by the Chl molecules associated with PSI, and at 562 nm, which is absorbed mainly by bile pigments and transferred to PSII, while the redox state of the cytochrome was monitored by the relative absorption of a weak measuring beam at 420 nm.

These observations are consistent with the series formulation or 'Z–scheme' of Hill and Bendall where the cytochrome is placed between the two photosystems. It is reduced when electrons are received from PSII and oxidized when they are passed on to PSI.

Since the Hill and Bendall scheme was proposed, a wide variety of experimental techniques has been used further to elucidate the way in which PSII and PSI are connected. The techniques have included: absorption difference spectrophotometry applied to cytochromes, quinones and the copper –containing protein plastocyanin (PC); extraction and re–addition procedures, particularly for quinones electron paramagnetic resonance (epr) spectroscopy, used for those compounds which do not show marked absorbance changes when oxidized or reduced inhibitor studies and studies on mutant algae and higher plants which show impaired electron transport.

Non–cyclic electron transport

Non–cyclic electron transport involves the flow of electrons from water, through PSII and PSI in series to the terminal electron acceptor NADP. The area of the electron transport chain between water and PSII is still in doubt, but Mn is probably involved, electron transport requires Cl^- and is inhibited by a number of treatments.

On the reducing side of PSII the first stable electron acceptor is Q, probably a quinone, which can apparently be detected through its absorbance at 320 nm. Recently it has become apparent that the pool of Q is composed of two components which can be distinguished on the basis of mid–point redox potential; the low potential component (Q_{lp}) has a midpoint redox potential of around – 250 mV while the high potential component (Q_{hp}), the previously designated 0, has a mid–point potential of approximately 0 mV. Q is a single electron acceptor; the next major component in the electron transport chain is plastoquinone (PQ) which accepts two electrons and at the same time takes up two protons. The reactions of plastoquinone in vitro have been studied recently by Rich.

In order for electron transport to proceed smoothly from 0 to PO there is an intermediate called R, also called B, which acts as an electron buffer, accepting two electrons from Q before handing them on to PQ. Plastoquinone also has a very important role to play in photophosphorylation.

Experiments using a mutant of the duckweed *Lemna* have indicated that electrons are passed from plastoquinone to a nonheme iron–sulfur (Fe–S) protein which is similar to that found in complex 3 of animal mitochondria, the so–called Rieske center. The Rieske center is unusual for an iron–sulfur protein, since they generally have rather negative redox potentials.

In contrast to cytochromes, the iron is held to the protein via cysteine–sulfur linkages; the sulfur is called labile, since it is released by mildly acidic conditions. The Rieske iron–sulfur center is oxidized by cytochrome f (Cyt –f) which is in its turn oxidized by plastocyanin (PC), a copper–containing protein which is the electron donor to P700.

Plastocyanin appears not to be strictly necessary: in some algae non–cyclic electron transport proceeds unimpeded although PC is absent. The reducing side of PSI is rich in iron–sulfur proteins: the electron acceptors X, A and B are all thought to be bound Fe–S centers.

The electrons then pass via ferredoxin– reducing substance (FRS) to a soluble Fe–S protein, ferredoxin (Fd), which is now well studied. The terminal electron acceptor is NADP which is reduced by ferredoxin via the ferredoxin–NADP reductase. The complete non-cyclic electron transport chain has been reviewed by Malkin.

At first sight it might be thought that the Z-scheme implies a fixed relationship between one PSII and one PSI that are rigidly linked by the electron transport chain. This is not so: it must be realized that the thylakoid membrane is fluid, and the PQ pool can make communication between several PSII and PSI possible. Moreover, chloroplasts may not contain equal numbers of PSII and PSI centers.

Cyclic electron transport

Electron transport does not only occur from water to NADP; there is a second 'mode' which involves neither the oxidation of water nor the reduction of NADP, and moreover only requires the excitation of photosystem 1. The pathway is called cyclic electron transport, and it was originally described by Arnon in his studies on photosynthetic phosphorylation.

Cyclic electron flow is not inhibited by 3(3,4–dichlorophenyl)1, 1 – dimethyl urea (DCMU), a powerful inhibitor of non –cyclic electron flow. In addition, it may involve a cytochrome (cyt–b_6) which does not appear to be involved in non–cyclic flow. In vitro cyclic electron flow normally requires the addition of cofactors to give appreciable rates.

Partial reactions

The isolation of chloroplasts normally yields a preparation that requires an exogenous electron acceptor before electron transport and the evolution of oxygen can proceed. The Hill oxidant potassium ferricyanide is one of the best known, but a variety of electron acceptors can be used which, depending on their mid–point redox potentials, will accept at different sites on the Z-scheme.

Similarly, if the oxidation of water is absent or inhibited, a suitable electron donor can be provided which will enable electron transport to proceed. In this way various sections of the electron transport chain ('partial reactions') can operate. For example, PSII alone mediates a partial reaction from water to the oxidized dye dichlorophenolin-dophenol (DCIP) which is reduced to its colorless form; PSI alone mediates electron transport from reduced DCIP to a PSI acceptor such as methyl viologen.

While the evolution of oxygen is normally inhibited in the latter reaction, the oxygen electrode may still be used to detect electron transport since the reduction of methyl viologen leads to the uptake of oxygen in a process known as the Mehler reaction.

A powerful technique in research into photosynthesis is the use of specific inhibitors of the electron transport chain. The best known and most widely used is DCMU, which inhibits electron transport between Q and plastoquinone. Commercially DCMU is used as a herbicide (diuron), and a good deal of attention has been paid to the protein on the chloroplast membrane to which it binds. The binding protein can be modified by the enzyme trypsin, which can then render electron transport DCMU– insensitive.

On the oxidizing side of PSII, electron transport can be inhibited by hydroxylamine, high concentrations of ammonia or methylamine, washing with high concentrations of the pH–buffer Tris, the uncoupler carbonyl cyanide 3–chlorophenyl hydrazone (CCCP), or treatment with a number of heavy metals.

Between the two photosystems the plastoquinone analog 2,5–dibromo–3–methyl–6–isopropyl–p–benzoquinone (dibromothymoquinone or DBMIB) inhibits the oxidation of plastoquinone, while plastocyanin is inhibited by poly–L–lysine, KCN or mercury. The reducing side of PSI is inhibited by disalicylidenpropane diamine (DSPD) and its water–soluble analog sulfo–DSPD.

Primary photoreactions

All reaction centers contain one or two molecules of Chi that are found in a special environment which makes the Chi molecules able to use very rapidly the excitation energy for photochemical work before it can move to the next light– harvesting Chi molecule.

The reaction center chlorophyll therefore constitutes a trap for excitation energy. When the energy is captured by the reaction center chlorophyll, it is used to put the chlorophyll into a state in which giving away an electron is a favored process; the structure of the center ensures that there is a suitable electron acceptor held very close to the reaction center chlorophyll, and which can take the electron, leaving the chlorophyll oxidized.

The spatial arrangement of the reaction center is also organized so that there is an electron donor in close proximity which can give an electron to the chlorophyll and return it to its former state. Designating the electron acceptor by A, the donor by D and the reaction center chlorophyll by P, the sequence is then

$$DPA + \text{light energy} \rightarrow DP^{*}A \rightarrow DP^{+}A^{-} \rightarrow D^{+}PA^{-}$$

Here P* represents the excited state of the chlorophyll.In purple bacteria, the ability of biochemists to isolate and purify reaction center complexes, and the ability of physicists to build light sources and detectors for spectrophotometers with picosecond time resolution has led to a good understanding of the details of charge separation in these organisms.

In green plants, however, the biochemistry is more difficult and consequently our knowledge is less advanced. Nevertheless, there is a good deal of' evidence which gives clues to how the reaction centers work.

Photosystem I

Photosystem I is characterized by a small reversible light–induced bleaching (decrease in absorbance) at 700 nm, first observed by Kok. The absorption difference spectrum has the characteristics of, chlorophyll, and the pigment responsible for it was named P – 700. The bleaching was associated with an electron paramagnetic resonance (epr) signal, which indicates the appearance of an unpaired electron, and suggests that the optical bleaching was associated with photo –oxidation. This result was confirmed by use of chemical oxidants.

Careful study of the linewidth of the P700 epr signal indicates that it is consistent with P700 being a pair of Chl–a molecules, which are so closely aligned in space that functionally they act as one single molecule. However this interpretation is not unequivocal. The nature of the substance that accepts the electron when P700 is oxidized to $P700^+$ has been studied mainly by epr spectroscopy since optical spectroscopy has not proved a fruitful technique. The primary acceptor may also be a Chl molecule, and the first stable acceptor is a bound iron–sulfur center denoted X; the secondary acceptors appear to be two further iron–sulfur centers (A and B) close by.

Photosystem II

Photosystem II are easier to study than those of photosystem I because a wider range of experimental techniques can be applied. But in spite of this apparent advantage some of the reactions of photosystem II, particularly those associated with the oxidation of water, remain enigmatic. Like PSI, PSII exhibits a small, reversible light-induced bleaching of a chlorophyll molecule, but in this case the maximum of the bleaching is at 680 rim, and hence the reaction–center pigment of PSII is termed P680.

Several epr signals are associated with PSII; the majority appear to be related to the oxidizing side of the reaction center. The components of the reaction center, other than P680, have been difficult to identify, but it has been suggested that the primary electron acceptor is a pheophytin molecule, that is, a Chi molecule whose central Mg has been replaced by two H^+.

Two phenomena associated with PSII are the emission of fluorescence from the light–harvesting chlorophylls and the oxidation of water. These will be described in turn.

Chlorophyll fluorescence

Chlorophyll fluorescence spectroscopy has considerably aided the study of PSII. It is very easy to measure, requiring only simple apparatus. Perhaps as a consequence of this simplicity the literature contains many reports on chlorophyll fluorescence: it is worth bearing in mind that even under the simplest circumstances a large number of factors come into play and influence the deceptively simple measurements.

Some of the energy absorbed by the chlorophyll molecules of PSII can be lost again as fluorescence. At room temperature in unfractionated chloroplasts, the emission is largely restricted to photosystem 11; the findings of Duysens and Sweers suggested that the yield of fluorescence (equal to the intensity of fluorescence divided by the intensity of illumination) was under the control of the redox state of the electron acceptor Q.

When Q was oxidized the fluorescence was quenched, but when Q was reduced the quenching was relieved. Indeed, Q stands for 'quencher'. The action of Q in quenching fluorescence can be readily understood in terms of the possible fates for excitation energy in photosystem II; these are

(i) re–emission as fluorescence,

(ii) radiationless decay to heat,

(iii) transfer to the weakly fluorescent PSI, and

(iv) use in photochemistry.

Giving these processes the rate constants K_f, K_h, K_t and K_p respectively, then the yield of fluorescence φ_f is given by

$$\varphi_f = \frac{K_f}{K_f + K_h + K_t + K_p A}$$

Here A is the fraction of photosystem II reaction centers that can accept an excitation, that is, centers where Q is oxidized. K_p is a relatively large rate constant when compared with the other rate constants, so φ_f is typically around 0.05. But if A decreases (more Q becomes reduced), φ_f will increase, because the denominator of the equation becomes smaller.

Conversely, if A increases (more Q becomes oxidized), φ_f will decrease; hence fluorescence is an inverse measure of the yield of photochemistry. There are two extreme cases, when $A = 1$ and $A = 0$. These give

$$\varphi_o = \frac{K_f}{K_f + K_h + K_t + K_p} \quad \text{and} \quad \varphi_m = \frac{K_f}{K_f + K_h + K_t}$$

If chloroplasts are dark–adapted so that all Q becomes oxidized ($A = 1$) and then illuminated in the presence of DCMU (to prevent reduced Q becoming re–oxidized by PQ), a fluorescence induction from a low to a high level is seen. Qualitatively, the Duysens and Sweer's hypothesis accounts quite well for the low and high levels.

While chlorophyll fluorescence is strictly speaking a property of PSII, the phenomenon can give information on electron transport if measurements are made in the absence of DCMU. The chlorophyll fluorescence induction under such circumstances is called the Kautsky effect. The redox state of Q certainly plays a role in determining the fluorescence yield, but in addition a more subtle factor comes into play. The absorption spectra of PSI and PSII are not identical, and consequently it is possible to illuminate a photosynthetic system so that either PSI or PSII is over–excited, and the under–excited photosystem becomes the rate–limiting step overall. This is the origin of the 'red drop' described above.

It now appears that the rate constant K_t for excitation energy transfer between PSII and PSI is not fixed, but can vary in order to equalize the excitation to the two photosystems in a process known as 'spillover'. It seems likely that the complex time course of the Kautsky effect is a mixture of spillover and quenching by Q, although there is now evidence that even the carbon dioxide fixation cycle may have an influence.

Chlorophyll fluorescence should be distinguished from another PSII light emission process, called delayed light emission, which is also known as luminescence or delayed fluorescence. This emission, which has an emission spectrum similar to that of 'prompt' fluorescence, is only seen after the exciting light has been extinguished.

It was first observed by Strehler and Arnold, and is thought to arise as a back–reaction between metastable stored intermediates in the reaction center of PSII. The yield of emission is very low. In principle it offers a probe into the very heart of PSII, but the intensity of the emission can be modulated by a number of factors, and in practice the phenomenon is very complex.

Oxygen evolution

The evolution of oxygen from water is one of the most important reactions of photosynthesis, and yet it remains one of the least understood. A glance at the stoichiometry indicates that the oxidation of water is chemically not an easy task.

$$2H_2O \rightarrow O_2 + 4H^+ + 4e^-$$

Since four electrons must be withdrawn from two molecules of water before molecular oxygen can be evolved. PSII can only withdraw one electron at a time: there are then two possibilities, either (i) four PSII reaction centers co–operate and pool their oxidizing equivalents to oxidize two molecules of water, or (ii) each PSII reaction center works independently and has a means of storing oxidizing equivalents until four are collected and can be used together. The first experimental data pertinent to distinguishing between these alternatives were obtained by Allen and Franck who found, using a sensitive phosphorescence method that in fully dark-adapted algae no oxygen was evolved by illumination with a single flash.

Similarly Joliot found that the evolution of oxygen in dark–adapted material required at least two photoacts for 'activation'. But a full answer to the question of co–operation or independence awaited a technical development, that of a rapidly– responding and sensitive oxygen electrode. The new electrode differed from the more usual Clarke–type found in many laboratories, in that the photosynthetic material was placed directly on the platinum electrode, and was not separated from it by a thin membrane.

Using a train of brief (10 μs) saturating flashes, the oxygen evolution elicited by each flash in the train could be observed. In this case the unicellular green alga Chlorella has been thoroughly dark–adapted before the experiment; there is no oxygen evolved on the first flash, little on the second, but a maximal amount on the third. The sequence continues with a period 4 in flash number, and gradually damps out until a steady–state level is seen on each flash.

An interpretation of this phenomenon was offered by Kok, Forbush and McGloin, and incorporated the concept of charge storage by each center. Each PSII reaction center is imagined to have a charge storage mechanism which can assume five states: S_0, S_1, S_2, S_3 and S_4 where the subscript denotes the oxidizing equivalents (power to extract electrons) stored.

Each absorbed photon transforms one state to the next with equal efficiency ($S_n \rightarrow S_{n+1}$); S_4 exists only transiently and reverts to S_0 after oxidizing water. The scheme needs the additional hypothesis that in the dark S_0 and S_1 are stable, and that during dark adaption the higher S–states revert mainly to S_1 and partially to S_0. There is some evidence for double hits, obtained from changing the duration of the flashes, but misses have posed a serious problem of interpretation.

Lavorel suggests that the independence of centers is not quite as strict as originally conceived by Kok et al., and that the oxygen–evolving apparatus consists of a fixed photoreaction which generates an oxidizing equivalent, and a mobile protein which stores the oxidizing equivalents, Misses occur when the mobile storage protein cannot collect an oxidizing equivalent from the fixed photoreaction center.

The phenomenology of flash–induced oxygen evolution says nothing about the biochemical nature of the intermediates involved. It has been known for some time that Mn is implicated in the evolution of oxygen, although not all of the pool of chloroplast Mn is essential.

Similarly, Cl^- and bicarbonate are essential co–factors for oxygen evolution. But interest has centered on Mn, since the multiplicity of oxidation states it possesses makes it plausible that it is involved in charge storage. Some use has been made of the Mn epr signal, which shows six strong peaks when it is free insolution, but these peaks disappear when it is bound. Washing chloroplasts with high concentrations of the buffer Tris is well known to inhibit the evolution of oxygen; using an epr spectrometer, Blankenship and Sauer showed that the inhibition was correlated with the release of bound Mn.

Similarly treatment of chloroplasts with hydroxylamine (NH_2OH) or mild heat, both of which inhibit oxygen evolution, are associated with the release of Mn. This circumstantial evidence was very useful, and a further step forward was made by studying the amount of Mn that could be released by mild heat treatment after each flash in a series of flashes. The amount of heat–releasable Mn shows oscillations with flash number of period 4, suggesting that it is closely associated with the oxygen–evolving mechanisms.

A further problem in the mechanism of oxygen evolution lies in the release of protons along with the oxidation of water. It might be thought that the protons would be released altogether at the same time as oxygen is released, but this is not the case. The sequence of proton release with respect to flash number is tricky to determine, but there is growing agreement on the sequence 1, 0, 1, 2.

Biochemical investigations

The investigations described in the foregoing sections have resulted in a picture of the way that pigments and electron transport components operate together, and in particular suggest the existence in higher plants and algae of two photosystems each of which contains a reaction center and has a light–harvesting complex of molecules attached to it. It is a natural question to ask whether these functional units can be separated as physical units using biochemical techniques.

There are two aims: firstly to confirm the description derived from other experiments, and secondly to make a simpler experimental material for the study of certain phenomena. Since oxygen–evolving organisms and photosynthetic bacteria show rather different characteristics, they will be treated separately.

Oxygen–evolving organisms

Starting with isolated chloroplasts, a number of techniques may be used to derive preparations enriched in PSI or PSII. Boardman, Thorne and Anderson treated chloroplasts with the detergent digitonin and found that after differential centrifugation the lighter of the two fractions so obtained had enhanced PSI activity and depressed PSII activity when compared to the control, and was moreover depleted in Chl–*b*. The heavier fraction was enriched in PSII activity and Chl–*b* as compared to the control.

Subsequently it was shown that the heavy fraction was composed mainly of granal stacks and that the light fraction consisted primarily of stromal lamellae. Similar separation can be achieved using other detergents, for example Triton X– 100, or by sonication or by passage of samples through the French pressure cell. However, it can be seen that the sub–chloroplast particles obtained by these methods are not a pure preparation of one photosystem. Nevertheless, the separation provides confirmation that PSI and PSII exist.

Further purification must consist in removing the pigments and electron transport components that are extraneous to the reaction centers, and it is there that the problem lies. Many attempts have been made, but it has proved very difficult satisfactorily to fractionate the preparations further, mainly because of the instability of the extracts.

Nevertheless, progress has been made towards purifying preparations of PSII. The protein composition of green plant chloroplasts has also attracted much interest thylakoid membranes are solubilized using the detergent sodium dodecyl sulfate (SDS) and the proteins separated using polyacrylamide gel electrophoresis. The results indicate that most if not all chlorophyll exists as pigment–protein complexes and that there are at least six major pigment-proteins, although these probably represent only three distinct chlorophyll–protein complexes. These are:

(i) a light–harvesting chlorophyll– protein complex which contains Chi–b, but has no light–induced oxidation – reduction activity associated with it

(ii) a PSI complex, which contains P700, several forms of Chl–a (as few as 40 Chi molecules, which include long– wavelength – absorbing forms) and a number of polypeptides; and

(iii) a fraction showing some PSII activity, but not the oxidation of water.

These three components which have presently been identified may well not describe the experimental data for long, since Work is progressing rapidly in this field. Nevertheless, tentative models for the organization of two reaction center complexes and a light– harvesting complex have been proposed, including the 'tripartite' model of Butler.

In addition, there have been attempts to correlate the very striking small and large particles which are apparent on freeze –fracture electron micrographs of thylakoid membranes with functional photosynthetic complexes.

Photosynthetic bacteria in marked contrast to higher plants, the isolation of functional complexes from photosynthetic bacteria was achieved several years ago. The isolation procedure is normally to break the cells, treat the exposed membranes with detergent and then use protein separation techniques to isolate a specific complex.

Photochemical reaction centers devoid of light--harvesting pigments have been obtained from a number of species. The reaction center from the purple non–sulfur bacterium Rhodopseudomonas sphaeroides contains 4 molecules of BChl and 2 of bacteriopheophytin (like BChl, but the Mg replaced by two H^+), one or two molecules of ubiquinone and one atom of ferrous iron, together with three polypeptides of apparent molecular weights in the region of 28, 24, and 21 kD. Reaction centers of this type have been the subject of intense study, and the photochemical reactions occurring in them are now quite well understood.

In addition, the reaction center polypeptides are now the subject of sequencing and crystallographic investigations. In agreement with the concept of reaction centers and light– harvesting pigments, there are also purely light– harvesting pigment–protein complexes, the study of which is only now beginning bacterial light – harvesting pigment–protein complex has been crystallized and subjected to X–ray analysis.

Photophosphorylation

ATP is the universal cellular energy carrier. It is formed from adenosine diphosphate (ADP) and inorganic phosphate (Pi) in oxidative and photosynthetic phosphorylation, and hydrolyzed with the consequent release of energy in a multitude of cellular reactions. The fixation of carbon dioxide within the chloroplast requires ATP.

The experiments of Arnon, Allen and Whatley demonstrated that ATP could be generated by the chloroplast itself, and not, as had previously been proposed, by oxidative phosphorylation. The energy required to make the phosphate bond between ADP and Pi comes from energy lost in electron transport, whether oxidative or photosynthetic.

The energy is passed via an intermediate known as the 'high–energy state' and often denoted by a so –called 'squiggle' bond, '~': it is the nature of the high–energy state that distinguishes the proposed mechanisms of phosphorylation.

Chemical hypothesis of phosphorylation

This hypothesis, first proposed by Slater suggested that a covalent bond was formed between a component of the electron transport chain, A, and some other entity, sometimes written as X. At the start A is reduced by a reaction center, but on oxidation by the next member of the electron transport chain, B, the redox energy which is lost is captured in turning the normal bond to the '~' bond between A and X. The '~' bond is then transferred until it is the phosphate bond between ADP and Pi. In summary

$$AH_2 + X + B \rightarrow A \sim X + BH_2$$

$$A \sim X + P_i \rightarrow A + X \sim P_i$$

$$ADP + X \sim P_i \rightarrow ATP + X$$

This mechanism has been examined very closely since it does not predict transmembrane electrochemical potential gradients, nor does it give a role for closed vesicles in phosphorylation.

Chemiosmotic hypothesis of phosphorylation

Proposed by Mitchell, this hypothesis suggests that the high –energy state consists of a proton–concentration gradient and an electric field across the functional membrane. It is general, and applies equally well to photosynthetic and oxidative phosphorylation both in prokaryotes and eukaryotes. There are five main postulates which are here applied to the chloroplast:

(a) the reactions take place on the thylakoid membrane, which forms closed vesicles, and which is essentially impermeable to protons;

(b) electron transfer from water to NADP takes place via pure electron carriers (like cytochromes) and hydrogen atom (electron plus proton) carriers; these carriers are arranged vectorially in the membrane;

(c) electron transfer is obligatorily linked to the pumping of hydrogen ions from the stroma into the intrathylakoid space, leading to the build–up of a transmembrane proton concentration gradient (ΔpH) and a transmembrane field ($\Delta\psi$); in the chloroplast the build–up of ΔpH – is favoured over that of $\Delta\psi$ by movement of counter–ions; (d) the combination of ΔpH and $\Delta\psi$ constitutes a store of energy (the high–energy state) and tends to expel protons from the intrathylakoid space;

(e) there is a membrane –bound vectorial ATP–synthetase enzyme that takes the energy of the electrochemical potential gradient of protons and uses it to form ATP from ADP and P_i.

Conformational hypothesis of phosphorylation

The third type of hypothesis suggests that events which occur within the membrane itself are important, and that the transfer of energy from electron transport to ATP formation is via forced conformational changes in the membrane or in the enzymes concerned with phosphorylation.

One major advantage of the chemiosmotic hypothesis was that it was open to experimental testing in a number of ways. In general, it has received good support from experiment, but there remain a number of gray areas where the precise mechanism remains in doubt. One such area is the arrangement of the proton pump; another is the functioning of the ATP–synthetase. The experimental evidence will be taken for each part of the hypothesis in turn.

The proton pump

It is an essential part of the chemiosmotic hypothesis that electron transport from water to NADP and the translocation of protons from the stroma to the intrathylakoid space are obligatorily linked or coupled. That proton movements of this type take place in chloroplasts was demonstrated by Hind and Jagendorf, where the loss of protons from an essentially unbuffered medium was detected by a pH–electrode in the medium. The increase in pH soon reached a steady state; the proton movements were reversed in the dark, and inhibited by inhibitors of electron transport, as well as uncouplers. The relationship between the electron transport chain and the proton pump clearly depends on the spatial arrangement of the electron–transport components in the thylakoid membrane.

Evidence for the location of particular components can be grouped into three forms:

(i) antibody agglutination or chemical labeling studies;

(ii) the accessibility of components to exogenous electron donors or acceptors before and after disrupting the thylakoid membrane; and

(iii) indirect studies.

For example, interaction between the reaction center chlorophyll of PSI, P700, and exogenous plastocyanin occurs only after the thylakoid membrane has been disrupted, suggesting P700 is located at the interior of the thylakoid vesicle. Similarly, antibodies raised to the protein plastocyanin cross–react only after disruption of the thylakoid, also suggesting that this component is at the interior.

According to this scheme there are two modes for proton movements.

In the first, plastoquinone, which when reduced is an electron–plus–proton carrier, is reduced at the stromal side of the membrane and becomes protonated there. The reduced entity (POH_2) then diffuses to the inside of the membrane where it is oxidized, and the protons are released. This is a direct example of the obligatory coupling between electron transport and proton pumping.

In the second mode, protons are taken up from the stroma as NADP is reduced, and released into the intrathylakoid space as water is oxidized, leading to an apparent movement of protons across the thylakoid membrane which is still strictly coupled to electron transport.

The ratio of protons translocated across the thylakoid membrane to electrons passed down the electron–transfer chain (the H^+/e ratio) has been the subject of a number of studies. In most cases the value of H+/e has been found to be 1, but there are a significant number of reports of H+/e being 2.

The experiments are rather difficult to perform and interpret; more data are needed before the value of the H+/e ratio is finally settled.

The high–energy state

In the chemiosmotic hypothesis the high–energy state is composed of the proton concentration gradient and the transmembrane field. Taken together, these two factors form the electrochemical potential gradient of protons, and comprise a store of energy which can be used to drive reactions in the energy–requiring ('uphill') direction.

Thermodynamically the parameter used to describe such energy is Gibbs free energy, which is calculated from the difference in the electrochemical potentials on either side of the membrane, $\bar{\mu}_{H^+}$ and these are given by

$$\bar{\mu}_{H^+} = \bar{\mu}^*_{H^+} + RT \ln H^+ + FE$$

where $\bar{\mu}^*_{H^+}$ is a constant, R is the gas constant, T the absolute temperature, F the Faraday and E the electrical potential. Taking the difference in $\bar{\mu}_{H^+}$ between the stroma and intrathylakoid space yields

$$\Delta\bar{\mu}_{H^+} = 2.303\ RT\ \Delta pH + F\Delta\psi$$

where ΔpH is the pH gradient across the membrane, and Δψ is the electrical potential difference. $\Delta\bar{\mu}_{H^+}$ represents the maximum amount of free energy available to do work; the expression is usually modified by dividing through by F, and renaming $\bar{\mu}_{H^+}$ /F the proton motive force (pmf). Recognizing that 2.303 RT/F is 59 mV at 20^{0}C, then

$$\text{pmf} = 59\Delta pH + Dy\ \text{mV}$$

The value of this expression is that it indicates the equal utility of ΔpH or Δψ in storing energy; thermo-dynamically the pmf can be all ΔpH and no Δψ or vice versa, or some of each.

The estimation of ΔpH and Δψ can be quite tricky, and usually entails following the distribution of a radioactively labeled permeant weak acid or amine to measure ΔpH, and a labeled permeant ion in the case of Δψ. In this way, the pmf in chloroplasts was found to be mainly ΔpH, with the gradient attaining values of three pH units. An experimental confirmation that a transmembrane pH gradient is the driving force for photophos-phorylation in chloroplasts was provided by Jagendorf and Uribe.

These workers decided to eliminate the H^+-pumping part of the photophosphorylation system, and impose an artificial transmembrane pH gradient in the dark. They suspended chloroplasts in a medium of low pH, and after allowing them to equilibrate transferred them to a medium of high pH, containing ADP, Pi and Mg^{2+}.

In this way a pH gradient of the correct polarity and size was imposed across the thylakoid membrane, and ATP was formed. There is an important consequence of the buildup of a large pH gradient across the thylakoid membrane. If protons are to be pumped across the thylakoid membrane to achieve a concentration difference of around 1000–fold (ΔpH = 3) then such an accumulation of positively charged ions will be accompanied by a membrane potential which is so large that it will oppose further proton influx. This will be prevented only if there is movement of counter–ions (anions inwards or cations outwards) to reduce the field to a small fraction of what it otherwise would be.

Measurements of the fluxes of H^+ and other ions have been made, either with ion –sensitive electrodes, or by indirect methods and the results indicate that the movement of Mg^{2+} from the intra–thylakoid space to the stroma and Cl^- in the reverse direction may be important counter–ion fluxes. The measurement of Δψ, at least at short times after flash excitation, can be studied very conveniently in photosynthetic organisms using spectrophotometry.

The basis of the method is that under the influence of the transmembrane field the pigments embedded in the membrane undergo a small shift in their absorption spectra, known as the electrochromic bandshift.

In chloroplasts, the maximum absorption is found at 515 nm in the green region of the spectrum. On flash illumination the bandshift shows a very rapid onset (< 20 ns), concomitant with the separation of charge at the reaction centers, followed by a slow decay whose half–time (20–200ms) is related to the decay of the transmembrane field due to ion movement through the membrane.

In steady–state illumination the absorption changes are much more complex because of light–induced scattering changes. Careful study of the kinetics of the onset of the electrochromic bandshift has revealed that under some circumstances the initial rapid absorption increase is followed by subsequent slower increase (half–time 5–20ms) that may represent the electrogenic movement of charge across the membrane that occurs at sites other than the reaction centers. Observations of this type have stimulated the interest in the existence of proton motive Q–cycles in chloroplasts that has been mentioned above.

The ATP–synthetase

The vectorial enzyme which catalyzes the formation of ATP at the expense of the proton–motive force is complex: it consists of a membrane – bound proton– conducting channel termed CF_0 (coupling factor 0) and a more loosely attached catalytic section called CF_1 (coupling factor 1). If the CF_1 is removed from the membrane either by brief sonication or by incubation with ethylene diamine tetra–acetic acid (EDTA; chelates divalent cations) the phosphorylation activity of the chloroplasts is lost and the thylakoids become leaky to H^+, as judged by H^+ uptake.

In electron micrographs of suitably stained material the CF_1 protein appears as a sphere of diameter about 9 nm; it has an apparent molecular weight of 325 000 daltons. After separation from the membrane CF_1 can be reconstituted and some M^{2+}–dependent phosphorylation activity is restored. CF_1 has latent ATP hydrolysis activity which can be activated with trypsin, and which is Ca^{2+}–dependent when the enzyme is removed from the membrane. The question of how many protons efflux through the ATP–synthetase per ATP formed (the H^+/ATP ratio) can be approached in two ways.

The first approach is thermodynamic, and seeks to equate the energy lost by protons as they pass through the ATP–synthetase to the energy required to phosphorylate ADP to ATP. Suitable calculations indicate that a H^+/ATP ratio of 3 is required to account for the synthesis of ATP found under circumstances when the pmf was known. Secondly, the approach can be experimental, when H^+/ATP ratios between 2 and 4 have been found! CF_1 is composed of five subunits whose molecular weights and subunit functions are now becoming clear.

One technique to determine subunit function has been to isolate and purify each subunit and to raise antibodies to it, then to challenge the intact coupling factor with the antibodies and to see which functions are inhibited. There has been a good deal of study on the binding of nucleotides to the CF_1 and on the influence of the high–energy state on such binding, as well as the concept of conformational changes generally.

During catalysis conformational changes on the enzyme take place, detected by the energy-dependent inclusion of tritiated water into the enzyme. The soluble CF_1 contains several nucleotide–binding sites: thoroughly washed CF_1 contains 2 ATP and 1 ADP per CF_1 which exchange only slowly in the dark, but exchange in the light is more rapid and uncouplersensitive.

An elegant reconstitution study was performed by Racker and Stockenius where a light–induced proton pump and the ATPase from beef heart were inserted into artificial phospholipid vesicles. The clever part of the experiment was the choice of the proton pump; rather than try to extract and reconstitute the complex proton pumps of the chloroplast or mitochondrion, Racker and Stockenius chose the simplest possible proton pump which is the 'purple patch' (a single protein called bacteriorhodopsin which pumps protons when illuminated) from a halophylic bacterium *Halobacterium halobium*. The purple patches and ATPase were reconstituted into the vesicles which were then illuminated, and ATP formation was found, suggesting that physical links between proton pump and ATPase were not necessary.

Uncouplers and ionophores

There are three separate places where photophosphorylation may be inhibited. The first place is the proton pump which, because it is obligatorily linked to electron transport, may be inhibited by the type of electron –transport inhibitor that has already been described. Secondly, inhibition can take place at the ATP synthesizing enzyme which can either be removed from the membrane by EDTA treatment or be blocked by substances called energy–transfer inhibitors.

The chemiosmotic hypothesis requires that the functional vesicle be impermeable to protons but a number of substances can artificially increase the proton permeability and hence wastefully dissipate the pmf.

Since this breaks the link between electron transport and phosphorylation, these compounds are called 'uncouplers'. Classically, they are lipophilic weak acids which can traverse the membrane in either protonated or unprotonated form, and hence dissipate both the proton concentration gradient and the transmembrane field.

Table 2
The subunits of the ATP-synthetase enzyme

Designation	*Molecular weight*	*Possible stoichio-metry*	*Suggested function*
α	59 700	2	High affinity nucleotide-binding site; regulation
β	55 700	2	Low affinity nucleotide-binding site; active site
γ	36 800	1	Energy transduction from pmf to ATP
δ	19 400	1	Attachment of CF_1 to the membrane
ε	14 100	2	Inhibits ATP-hydrolase activity

Some examples are dinitrophenol (DNP), carbonyl cyanide–*p*–trifluoromethoxyphenylhydrazone (FCCP), and CCCP which has been mentioned above. There also exist ionophores which form lipophilic ionophoreion complexes and hence render membranes permeable to specific ions, e.g., valinomycin is a specific K^+ carrier; nigericin catalyzes the neutral exchange of K^+ for H^+.

Fixation of carbon dioxide

The key experimental tools to solve the problem of working out the biochemical pathway for the incorporation of carbon dioxide into sugar were firstly the availability of the stable radioactive isotope of carbon, ^{14}C, and secondly the ability to separate similar substances by paper chromatography.

With these tools, Calvin and his colleagues working in Berkeley in the made great strides in the understanding of the dark reactions of photosynthesis. The experiments consisted in illuminating algae in the presence of $H^{14}CO_3^-$ and after various times killing and extracting the algae in hot alcohol.

The extract was then separated using two–dimensional chromatography, and radioactive compounds were identified by autoradiography.

Calvin cycle (C–3 cycle)

Calvin and his co–workers found a general pattern in their experiments the first radioactively labeled compound was phosphoglycerate (PGA) which contains 3 carbon atoms–hence the name 'C–3 cycle'. This was produced as a result of the carboxylation of the 5–carbon compound ribulose–1.5–bisphosphate (RuBP), and the reaction was catalyzed by the most abundant protein in the world, ribulose–1,5–bisphosphate carboxylase (reaction 1).

$$3RuBP + 3CO_2 + 3H_2O \rightarrow 6PGA$$

After carboxylation, PGA is phosphorylated to diphosphoglycerate (1,3–DPGA) at the expense of a molecule of ATP.

$$6\ PGA + 6\ ATP \rightarrow 6\ (1,3 - DPGA) + 6\ ADP$$

The 1,3–DPGA is then reduced with NADPH and a proton to glyceraldehyde – 3 – phosphate (G3P), which can be regarded as the product of the cycle (reaction III)

$$6\ (1,3\text{–}DPGA) + 6\ NADPH + 6\ H^+ \rightarrow 6\ G3P + 6\ NADP+ + 6\ P_i$$

Looked at in its simplest terms, the cycle involves four reactions: carboxylation of RuBP to 2 PGA, phosphorylation of PGA to 1,3–DPGA, reduction of 1,3–DPGA to G3P and lastly the regeneration of RuBP from

G3P, a reaction which also calls for chemical energy derived from ATP. The net equation of the Calvin cycle can be written

$$3CO_2 + 9ATP + 6NADPH + 6H^+$$
$$\rightarrow G3P\ (Product) + 9\ ADP + 6\ NADP^+ + 8\ P_i$$

Now that the Calvin cycle itself is well established, attention has more recently been centered on the regulation of the cycle and on the properties of its enzymes. Regulation of the Calvin cycle has been discussed in detail by Edwards and Walker. It is important because both oxidative and reductive carbohydrate metabolism occur in the same organelle.

If the Calvin cycle were not de–activated in the dark there is the possibility that ATP would be wasted in 'futile cycles', one of which might be the continuous inter–conversion of fructose–6–phosphate and fructose– 1,6–diphosphate catalyzed by the enzymes phosphofructokinase and fructose–1,6–diphosphatase. Each turn of the cycle would wastefully consume one molecule of ATP.

The problem of such a 'futile cycle' could be resolved if one of the enzymes were regulated, so that both were not simultaneously catalytic. Activation of some enzymes of the Calvin cycle by light has been known for some time. In particular, the enzymes fructose–1,6-diphosphatase, sedoheptulose–1,7-diphosphatase, NADP–glyceraldehyde–3–phosphate dehydrogenase and phosphoribulokinase are activated by light in chloroplasts in a DCMU–sensitive manner. This activation can also be achieved outside the chloroplast by reducing agents such as dithiothreitol, and has led to the suggestion that activation in vivo is due to reducing agents which are formed in the light as a result of non–cyclic electron transport.

The hypothesis has two variations; the first is proposed by Buchanan and suggests that small soluble proteins, named thioredoxin and ferredoxin – thioredoxin reductase (FTR) participate. Thioredoxin is thought to be reduced by ferredoxin via FTR, and reduced thioredoxin activates the sensitive enzyme by reducing disulfide groups on the enzyme to sulfhydryl groups.

The second variation has been proposed by Anderson and co–workers, and suggests the involvement of a membrane –bound component (the 'light effect mediator', LEM) which is reduced directly by the electron–transport chain, and then alters the conformation of the sensitive enzyme.

An alternative hypothesis for the regulation of the Calvin cycle is linked to the idea of proton movements from the stroma into the intra–thylakoid space being coupled with the movement of Mg^{2+} in the opposite direction to maintain electroneutrality. Since the chloroplast envelope is relatively impermeable to ions, the movements of H^+ and Mg^{2+} imply that their concentrations in the stroma will vary; both fructose– 1,6–diphosphatase and sedoheptulose–1,6–diphosphatase are sensitive to pH and Mg^{2+} concentration and it has been suggested that the movement of these ions in response to electron transport may switch on the Calvin cycle in the light.

C–4 cycle

The labeling pattern ^{14}C found with the Calvin cycle is not found in all plants. In some, the first radioactively labeled compound contains 4 carbon atoms (e.g., malate, aspartate) and plants showing this character are termed 'C–4 plants'. They have a distinguishing anatomical feature as well: the presence of a layer of bundle –sheath cells containing chloroplasts around the vascular tissue in the leaf. This feature is called Kranz anatomy, and is not seen in C–3 plants.

Moreover, there are differences in chloroplast ultrastructure between the mesophyll and bundle–sheath cells; mesophyll cell chloroplasts have normal thylakoid membranes which are stacked into grana in some regions, but lack starch grains, whereas bundle–sheath cell chloroplasts are agranal but have starch grains. There are, however, subtle differences in the ultrastructure of the bundle–sheath cell chloroplasts between the different types of C–4 plants. The distinguishing biochemical feature of C–4 plants is the original carboxylation of CO_2. The CO_2 acceptor is the 3–carbon compound phosphoenolpyruvate (PEP), and the product is the 4–carbon compound oxaloacetate (OAA). The reaction occurs in the mesophyll cells.

The fate of the OAA is of the same general pattern in all C–4 plants, but varies in the detail. Taking maize (*Zea mays*) as an example, the OAA is reduced to malate at the expense of reduced NADP; the malate so formed is then exported to the bundle–sheath cells where it is decarboxylated to pyruvate, and the CO_2 released is carboxylated by RuBP carboxylase into PGA.

The pyruvate returns to the mesophyll cells where it is converted to PEP at the expense of ATP. It should be noted that the C–4 cycle does not achieve any net carboxylation; it merely supplies carbon dioxide to the C – 3 cycle via a different route. It costs metabolic energy (extra ATP per CO_2), but since the K_m for CO_2 of PEP carboxylase is lower than that of RuBP carboxylase, the cycle can act as a CO_2– concentrating device. C–4 plants can be divided into three types, depending on the details of the intermediates of the cycle.

There are two types of aspartate former, both of which form aspartate from OAA in mesophyll cells and then transfer it to the bundle–sheath cells, where the difference between the two types becomes apparent.

In aspartate formers of type 1, aspartate is converted back to OAA, which is then decarboxylated to PEP using a molecule of ATP as it does so. The fate of the PEP is not clear: it may be returned to the mesophyll cells, or it may participate in the return of nitrogen to the mesophyll cells.

Aspartate formers of type 2 transfer aspartate to the bundle-sheath cells as described above, but the formation of OAA is followed by that of malate, which is then decarboxylated to pyruvate. The pyruvate is transaminated to alanine, which is the route by which nitrogen returns to the mesophyll cells.

The three types of C–4 plants can equally well be named according to the predominant enzyme which catalyzes the decarboxylation of the C–4 compound in the bundle–sheath cells. Malate formers have the NADP–dependent malic enzyme (NADP–ME), aspartate formers of type I have phosphoenolpyruvate carboxykinase (PEP–CK) and aspartate formers of type 2 have the NAD–dependent malic enzyme (NAD–ME).

Edwards and Huber have summarized the features of each type of C–4 plant; it should be realized that the relative amount of ^{14}C found in a C–4 product is to some extent dependent on environmental and other factors.

Crassulacean acid metabolism

Crassulacean acid metabolism (CAM) is an interesting variation of carbon metabolism that occurs in succulent plants that normally inhabit dry conditions. During the night the stomata are open, and the level of malic acid in the vacuole increases, whereas during the day the stomata close and the level of malic acid falls. Carbon dioxide uptake from the atmosphere occurs predominantly during darkness.

CAM plants contain the enzymes of the C–4 cycle, and it appears that carboxylation of PEP to malate occurs during the night when the stomata are open, and a pool of malate is formed. During the day the stomata are closed to reduce water loss, and the malate Is decarboxylated to provide a source of CO_2 for the Calvin cycle. Hence, what the C–4 plant separates in space, the CAM plant separates in time. CAM appears to be an adaptation to reduce water loss during the day in dry habitats, and yet to maintain photosynthesis.

Chapter 9

Biochemistry Protein Synthesis

Structure of proteins

Many proteins have no obvious regular structure, but this is deceptive and each protein has a precise structure needed for its proper function. As a consequence, proteins are frequently described as having large segments of random structure. The term random is really a misnomer, since the same complex structure is found in all molecules of a given protein having the same native conformation. There are mainly four types of structural levels in proteins.

1) *Primary structure*: Primary structure is the order in which the amino acids are linked together. The peptide Leu-Gly-Thr-Val-Arg-Asp-His has a different primary structure from the peptide Val-His-Asp-Leu-Gly-Arg-Thr, even though both have the same number and kinds of amino acids. The order of amino acids can be written on one line.

2) *Secondary structure*: Secondary structure is the arrangement in space of the atoms in the backbone of the polypeptide chain. The a-helix and b-pleated sheet hydrogen-bonded arrangements are two different types of secondary structure. In very large proteins, the folding of parts of the chain can occur independently of the folding of other parts. Such

independently folded portions of proteins are referred to as domains or supersecondary structure.

3) *Tertiary structure*: Tertiary structure includes the three-dimensional arrangement of all the atoms in the protein, including those in the side chains and in any prosthetic groups. A protein can consist of multiple polypeptide chains called subunits.

4) The arrangement of subunits with respect to one another is the quaternary structure. Interaction between subunits is mediated by noncovalent interactions, such as hydrogen bonds and hydrophobic interactions.

Primary structure

The primary structure of a protein determines its three-dimensional structure, which in turn determines its properties. In every protein the correct three-dimensional structure is needed for correct functioning. One of the most striking demonstrations of the importance of primary structure is found in the haemoglobin associated with sickle-cell anaemia.

In this genetic disease, red blood cells cannot bind oxygen efficiently. The red cells also assume a characteristic sickly shape, giving the disease its name. The sickled cells tend to become trapped in small blood vessels, cutting off circulation and thereby causing organ damage. These drastic consequences stem from a change in one amino acid residue in the sequence of the primary structure.

Considerable research is being done to determine the effects of changes in primary structure on the functions of proteins. It is possible to replace any chosen amino acid residue in a protein with another specific amino acid residue. The conformation of the altered protein, as well as its biological activity, can then be determined. The results of such amino acid substitutions range from negligible effects to complete loss of activity, depending on the protein and the nature of the altered residue.

Sequencing of peptides

The sequence of amino acids determination consists of several steps, which must be carried out carefully to obtain accurate results.

In first step in determining the primary structure of a protein is to establish which amino acids are present and in what proportions. It is relatively easy to break a protein down to its component amino acids by heating a solution of the protein in acid, usually 6 M HCI, at 100°C to 110°C for 12 to 36 hours to hydrolyze the peptide bonds. Separation and identification of the products are somewhat more difficult and are best done by an automated instrument known as an amino acid analyser. This instrument gives both qualitative information about the identities of the amino acids present and quantitative information about the relative amounts of those amino acids.

The second step identities of the N-terminal and C-terminal amino acids in a protein sequence are determined. This procedure is becoming less and less necessary as a separate step as the sequencing of individual peptides improves, but it can be used to check whether a protein consists of one or two polypeptide chains. When the end-group analysis is complete, it is possible to proceed with steps three and four, the determination of the amino acid sequence.

Automated instruments can perform a stepwise modification starting from the N-terminal end, followed by cleavage in turn of each amino acid in the sequence, and subsequent identification of each modified amino acid as it is removed. The process becomes more difficult as the number of amino acids increases. It is usually necessary to break a long polypeptide chain into fragments of reasonable size for sequencing.

Proteins can be cleaved by enzymes or by chemical reagents at specific sites. The enzyme trypsin cleaves peptide bonds preferentially at amino acids that have positively charged R groups, such as lysine and arginine. The cleavage takes place in such a way that the amino acid with the charged side chain ends up at the C-terminal end of one of the peptides produced by the reaction.

The C-terminal amino acid of the original protein can be any one of the 20 amino acids and is not necessarily one at which cleavage takes place. A peptide can be automatically identified as the C-terminal end of the original chain if its C-terminal amino acid is not a site of cleavage. Another enzyme, chymotrypsin, cleaves pepticle bonds preferentially at aromatic-amino acids such as tyrosine, tryptophan, and phenylalanine.

The aromatic amino acid ends up at the C-terminal ends of the peptides produced by the reaction. In the case of the chemical reagent cyanogen bromide (CN-Br), the sites of cleavage are at internal methionine residues. The sulphur of the methionine reacts with the carbon of the cyanogen bromide to produce a homoserine lactone at the C-terminal end of the fragment.

The cleavage of a protein by any of these reagents produces a mixture of peptides, which can be separated by standard methods. The use of several such reagents on different samples of a protein to be sequenced produces different mixtures. The sequences of a set of peptides produced by one reagent will overlap the sequences produced by another reagent. This makes it possible to arrange the peptides in the proper order once their own sequences have been determined.

Sequencing of each peptide produced by specific cleavage of a protein is accomplished by repeated application of a procedure called Edman degradation. The sequence of a peptide containing 10 to 20 residues can easily be determined by this method in about 30 minutes using as little as 10 picomoles of material. It then becomes possible to determine the sequence of the entire protein.

The overlapping sequences of peptides produced by different reagents provide the key to solving the puzzle. The alignment of like sequences on different peptides makes it possible to deduce the overall sequence. The Edman method has become so efficient that it is no longer considered necessary to identify the N-terminal and C-terminal ends of a protein by chemical or enzymatic methods. It is necessary to keep in mind, however, that a protein may consist of more than one polypeptide chain and to keep this point in mind in interpreting results.

In the sequencing of a peptide, the Edman reagent, phenyl isothiocyanate, reacts with peptide's N-terminal residue. The modified amino acid can be cleaved off, leaving the rest of the peptide intact, and can be detected as the phenylthiohydantoin derivative of the amino acid. The second amino acid of the original peptide can then be treated in the same way, and then the third.

With an automated instrument called a sequencer, the process is repeated until the whole peptide is sequenced. Another sequencing method makes use of the fact that the amino acid sequence of a protein reflects the base sequence of the DNA in the gene that coded for that protein.

Secondary structure

The secondary structure of protiens is the hydrogen bonded arrangement of backbone protein, the polypeptide chain. The nature of the bonds in peptide backbone plays an important role here. Within each amino acid residue are two bonds with reasonably free rotation. They are

i) the bond between the α-carbon and the amino nitrogen of that residue and

ii) the bond between the α-carbon and the carboxyl carbon of that residue.

The combination of the planar peptide group and the two freely rotating bonds has important implications for the three-dimensional conformations of peptides and proteins. A peptide chain backbone can be visualized as a series of playing cards, each card representing a planar peptide group.

The cards are linked at opposite corners by swivels, representing the bonds about which there is considerable freedom of rotation. The side chains also play a vital role in determining the three-dimensional shape of a protein, but only the backbone is considered in secondary structure. Angles ϕ and ψ, frequently called Ramachandran angles after their originator G.N. Ramachandran, are used to designate rotations around the C-N and C-C bonds, respectively.

The conformation of a protein backbone can be described by specifying the values of ϕ and ψ for each residue. Two kinds of secondary structures that occur frequently in proteins are the repeating α-helix and β-pleated sheet hydrogen-bonded structures. The ϕ and ψ angles repeat themselves in contiguous amino acids in regular secondary structures.

The α-helix and β-pleated sheet are not the only possible secondary structures, but they are by far the most important, and they deserve a closer look. The α-helix and β-pleated sheet are periodic structures; their features repeat at regular intervals. The α-helix is rodlike and involves only one polypeptide chain. The β-pleated sheet structure can give a two-dimensional array and can involve one or more polypeptide chains.

The α-helix is stabilized by hydrogen bonds parallel to the helix axis within the backbone of a single polypeptide chain. Counting from the N-terminal end, the C=O group of each amino acid residue is hydrogen bonded to the N-H group of the amino acid four residues away from it in the covalently bonded sequence.

The helical conformation allows a linear arrangement of the atoms involved in the hydrogen bonds, which gives the bonds maximum strength and thus makes the helical conformation very stable. There are 3.6 residues for each turn of the helix, and the pitch of the helix (the linear distance between corresponding points on successive turns) is 5.4 Å. Proteins have varying amounts of α-helical structures, varying from a few percent to nearly 100%.

Several factors can disrupt the α-helix. The amino acid proline creates a bend in the backbone because of its cyclic structure. It cannot fit into the α-helix for two reasons: i) rotation around the bond between the nitrogen and the cc-carbon is severely restricted, and ii) proline's cc-amino group cannot participate in intrachain hydrogen bonding. Other localized factors involving the side chains include strong electrostatic repulsion caused by the proximity of several charged groups of the same sign, such as groups of positively charged lysine and arginine residues or groups of negatively charged glutamate and aspartate residues.

Another possibility is crowding caused by the proximity of several bulky, side chains. In the cc-helical conformation, all the side chains lie outside the helix; there is not enough room for them in tile interior. The β–carbon is just outside the helix, and crowding can occur if it is bonded to two atoms other than hydrogen, as is the case with valine, isóleucine, and threonine. Arrangement of atoms in the β-pleated sheet conformation differs markedly from that in the α-helix. The peptide backbone in the β-sheet is almost completely extended.

Hydrogen bonds can be formed between different parts of a single chain that is doubled back on itself (intrachain bonds) or between different chains (interchain bonds). If the peptide chains run in the same direction (i.e., if they are all aligned in terms of their N-terminal and C-terminal ends), a parallel pleated sheet is formed. When alternating chains run in opposite directions, an antiparallel pleated sheet is formed.

The hydrogen bonding between peptide chains in the P-pleated sheet gives rise to a repeated zigzag structure—hence, the name pleated sheet. The α-helix and β-pleated sheet are combined in many ways as the polypeptide chain folds back on itself in a protein. For steric (spatial) reasons glycine is frequently encountered in reverse turns, at which the polypeptide chain changes direction; the single hydrogen of' the side chain prevents crowding.

Because the cyclic structure of proline has the correct geometry for a reverse turn, this amino acid is also frequently encountered in such turns. The combination of α- and β-strands produces various kinds of supersecondary structures in proteins. The most common feature of this sort is the βαβ unit, in which two parallel strands of β-sheet are connected by a stretch of α-helix. An αα unit consists of two antiparallel α-helices.

In such an arrangement, energetically favourable contacts exist between the side chains in the two stretches of helix. In β-meander, an antiparallel sheet is formed by a series of tight reverse turns connecting stretches of the polypeptide chain. Another kind of antiparallel sheet is formed when the polypeptide chain doubles back on itself in a pattern known as the Greek key, named for a decorative design found on pottery from the classical period.

When β-sheets are extensive enough they can fold back on themselves, forming a β-barrel, a structural feature that occurs in many proteins. The nature of the side chains in a protein (part of the tertiary structure) can influence the folding of the backbone (the secondary structure). A comparison of collagen with silk and wool fibers can be illuminating. Silk fibers consist largely of the protein fibroin, which, like collagen, has a fibrous structure but which, unlike collagen, consists largely of β-sheets.

Fibers of wool consist largely of the protein keratin, which is largely α-helical. The amino acids of which collagen, fibroin, and keratin are composed determine which conformation they will adopt, but all are fibrous proteins. In other proteins, the backbone folds back on itself to produce a more or less spherical shape. These are called globular proteins, and we shall see many examples of them. Their helical and pleated sheet sections can be arranged so as to bring the ends of the sequence close to each other in three dimensions.

Tertiary structure

The conformations of the side chains and the positions of any prosthetic groups are parts of the tertiary structure, as is the arrangement of helical and pleated sheet sections with respect to one another. In a fibrous protein, the overall shape of which is a long rod, the secondary structure provides much of the information about the tertiary structure as well.

The helical backbone of the protein does not fold back on itself, and the only important aspect of the tertiary structure that is not specified by the secondary structure is the arrangement of the atoms of the side chains. For a globular protein, considerably more information is needed. It is necessary to determine the way in which the helical and pleated sheet sections fold back on each other, in addition to the positions of the side-chain atoms and any prosthetic groups.

The interactions between the side chains play an important role in the folding of proteins. The folding pattern frequently brings residues that are separated in the amino acid sequence into proximity in the tertiary structure of the native protein. Primary structure of a protein depends on the formation of peptide bonds, which are covalent.

Higher-order levels of structure, such as the conformation of the backbone (secondary structure) and the positions of all the atoms in the protein (tertiary structure), depend on noncovalent interactions; if the protein consists of several subunits, the interaction of the subunits (quaternary structure) also depends on noncovalent interactions. Noncovalent stabilizing forces contribute to the most stable structure for a given protein, the one with the lowest energy.

Several types of hydrogen bonding occur in proteins. Backbone, hydrogen bonding is a major determinant of secondary structure; hydrogen bonds between the side chains of amino acids are also possible in proteins. Nonpolar residues tend to cluster together in the interior of protein molecules as a result of hydraphobic interactions.

Electrostatic attraction between oppositely charged groups, which frequently occurs on the surface of the molecule, results in such groups being close to one another. Several side chains can be complexed to a single metal ion.

In addition to these noncovalent interactions, disulfide bonds form covalent links between the side chains of cyteines. When such bonds form, they restrict the folding patterns available to polypeptide chains. There are specialized laboratory methods for determining the number and positions of disulfide links in a given protein.

Information about the locations of disulfide links can then be combined with knowledge of the primary structure to give the complete covalent structure of the protein. Note the subtle difference here: The primary structure is the order of amino acids, whereas the complete covalent structure also specifies the positions of the disulfide bonds.

As a result of this assortment of stabilizing forces, residues that are far apart in the primary sequence can be close to each other in the three dimensional structure produced by the folding of the protein. When a polypeptide chain folds back on itself, it can assume a compact globular shape. A different polypeptide chain or the same chain under different conditions can assume a rodlike fibrous form. Not every protein necessarily exhibits all possible structural features.

For instance, there are no disulfide bridges in myoglobin and haemoglobin, which are oxygen storage and transport proteins and classic examples of protein structure, but they both contain Fe (II) ions as part of a prosthetic group. In contrast, the enzymes trypsin and chymotrypsin do not contain complexed metal ions, but they do have disulfide bridges.

Hydrogen bonds, electrostatic interactions, and hydrophobic interactions occur in most proteins. The three-dimensional conformation of a protein is the result of the interplay of all the stabilizing forces. It is known, for example, that proline does not fit into an α-helix and that its presence can cause a polypeptide chain to turn a corner, ending an α-helical segment.

The presence of proline is not, however, a requirement for a turn in a polypeptide chain. Other residues are routinely encountered at bends in polypeptide chains. The segments of proteins at bends in the polypeptide chain and in other portions of the protein that are not involved in helical or pleated sheet structures are the ones frequently referred to as random or random coil. Experimental technique used to determine the tertiary structure of a protein is x-ray crystallography. Perfect crystals of some proteins can be grown under carefully controlled conditions.

In such a crystal, all the individual protein molecules have the same three-dimensional conformation and the same orientation. Crystals of this quality can be formed only from proteins of very high purity, and it is not always possible to obtain a structure if the protein cannot be crystallized. When a suitably mounted protein crystal is exposed to a beam of x-rays, a diffraction pattern is produced on a photographic plate or a radiation counter. The pattern is produced as a result of scattering of the x-rays by the electrons in each atom in the molecule.

The number of electrons in the atom determines the intensity of its scattering of x-rays; heavier atoms scatter more effectively than light atoms. The scattered x-rays from the individual atoms can reinforce each other or cancel each other giving rise to the characteristic pattern for each type of molecule. A series of diffraction patterns taken from several angles contains the information needed to determine the tertiary structure.

The information is extracted from the diffraction patterns through a mathematical analysis known as a Fourier series. Many thousands of such calculations are required to determine the structure of a protein, and even though they are performed by computer, the process is a fairly long one. Improving the calculation procedure is a subject of active research. A form of nuclear magnetic resonance (NMR) spectroscopy which supplements the results of x-ray diffraction has come into wide use in recent years. In this particular application of NMR, called 2-D (two-dimensional) NMR, large collections of data points are subjected to computer analysis.

Like x-ray diffraction, this method uses a Fourier series as the mathematical tool for analysis of results. It is similar to x-ray diffraction in another way. It is a long process and requires considerable amounts of computing power. One way in which 2-D NMR differs from x-ray diffraction is that, rather than crystals, it uses protein samples in aqueous solution. This is closer to the state of proteins in cells, and thus it is one of the main advantages of the method.

The NMR method most widely used in protein structure determination ultimately depends on the distances between hydrogen atoms, giving results independent of those obtained by x-ray crystallography. Myoglobin was the first protein for which the complete tertiary structure was determined by x-ray crystallography. The complete myoglobin molecule consists of a single polypeptide chain of 153 amino acid residues and includes a prosthetic group, the heme group, which also occurs in haemoglobin.

The myoglobin molecule has a compact structure, with the interior atoms very close to each other. This structure provides examples of many of the forces responsible for the three-dimensional shapes of proteins. In myoglobin there are eight α-helical regions and no β-pleated sheet regions. Approximately 75 per cent of the residues in myoglobin are found in these helical regions, which are designated by the letters A though H. Hydrogen bonding in the polypeptide backbone stabilizes the α-helical regions; amino acid side chains are also involved in hydrogen bonds. The polar residues are on the exterior of the molecule.

The interior of the protein contains almost exclusively nonpolar amino acid residues. Two polar histidine residues are found in the interior; they are involved in interactions with the heme group and bound oxygen, and thus they play an important role in the function of the molecule.

The planar heme group fits into a hydrophobic pocket in the protein portion of the molecule and is held in position by hydrophobic attractions between heme's porphyrin ring and the nonpolar side chains of the protein. Presence of the heme group drastically affects the conformation of the polypeptide: The apoprotein is not as tightly folded as the complete molecule. The heme group consists of a metal ion, Fe(II), and an organic part, protoporphyrin IX. The porphyrin part consists of four five-membered rings based on the pyrrole structure; these four rings are linked by bridging methine (-CH=) groups to form a square planar structure. The Fe(II) ion has six coordination sites, and it forms six metal-ion complexation bonds. Four of the six sites are occupied by the nitrogen atoms of the four pyrrole-type rings of the porphyrin to give the complete heme group.

The presence of the heme group is required for myoglobin to bind oxygen. The fifth coordination site of the Fe(II) ion is occupied by one of the nitrogen atoms of the imidazole side chain of histidine residue F8 (the eighth residue in helical segment F). This histidine residue is one of the two in the interior of the molecule. The oxygen is bound at the sixth coordination site of the iron. The fifth and sixth coordination sites lie perpendicular to and on opposite sides of the plane of the porphyrin ring.

The other histidine residue in the interior of the molecule, residue E7 (the seventh residue in helical segment E), lies on the same side of the heme group as the bound oxygen. This second histidine is not bound to the iron or to any part of the heme group, but it acts as a gate that opens and closes as oxygen enters the hydrophobic pocket to bind to the heme. In the absence of the protein the iron of the heme group can be oxidized to Fe(III); the oxidized heme will not bind oxygen. The unfolding of a protein is called denaturation. Reduction of disulfide bonds leads to even more extensive unravelling of the tertiary structure.

Denaturation and reduction of disulfide bonds are frequently combined when complete disruption of the tertiary structure of proteins is desired. Under proper experimental conditions, the disrupted structure can then be completely recovered. This process of denaturation and refolding is a dramatic demonstration of the relationship between the primary structure of the protein and the forces that determine the tertiary structure. Proteins can be denatured by an increase in temperature favours vibrations within the molecule, and the energy of these vibrations can become great enough to disrupt the tertiary structure.

At either high or low extremes of pH, at least some of the charges on the protein are missing, and so the electrostatic interactions that would normally stabilize the native, active form of the protein are drastically reduced. This leads to denaturation. The binding of detergents such as sodium dodecyl sulphate (SDS) also denatures proteins. Detergents tend to disrupt hydrophobic interactions. If a detergent is charged, it can disrupt electrostatic interactions within the protein, as well. Other reagents such as urea and guanidine hydrochloride form hydrogen bonds to the protein that are stronger than those within the protein.

These two reagents can also disrupt hydrophobic interactions in much the same way as detergents. Mercaptoethanol ($HS—CH_2—CH_2—OH$) is frequently used to reduce disulfide bridges to two sulin blood sulphydryl groups. Urea is usually added to the reaction mixture to facilitate unfolding of the protein and to increase the accessibility of the disulphides to the reducing agent. If experimental conditions are properly chosen, the native conformation of the protein can be recovered when both mercaptoethanol and urea are removed.

Quaternary structure

Quaternary structure of proteins consist of more than one polypeptide chain. The number of chains can range from two to more than a dozen, and the chains may be identical or different. Commonly occurring examples are dimers, trimers, and tetramers, consisting of two, three, and four polypeptide chains, respectively. The chains interact with one another noncovalently. As a result of these noncovalent interactions, subtle changes in structure at one site on a protein molecule may cause drastic changes in properties at a distant site.

Proteins that exhibit this property are called allosteric. Not all multisubunit proteins exhibit allosteric effects, but many do. A classic illustration of quaternary structure of proteins and its effect on properties is a comparison of haemoglobin, an allosteric protein, with myoglobin, which consists of a single polypeptide chain. We can take hemoglobin as an example to tetramer. It consisting of four polypeptide chains, two α-chains, and two β-chains. The two α-chains of hemoglobin are identical, as are the two β-chains.

The overall structure of hemoglobin is $\alpha_2\beta_2$ in Greek letter notation. Both the α- and β-chains of hemoglobin are very similar to the myoglobin chain. The α-chain is 141 residues long and the β-chain is 146 residues long. Lengthy sequences of the α-chain, the β-chain, and myoglobin are homologous; that is, most or all of the amino acid residues are in the same positions. The heme group is the same in myoglobin and hemoglobin. Four molecules of oxygen can bind to one hemoglobin molecule.

Both hemoglobin and myoglobin bind oxygen reversibly, but the binding of oxygen to hemoglobin is cooperative whereas oxygen binding to myoglobin is not. Cooperative binding means that when one oxygen molecule is bound, it becomes easier for the next to bind. When the degree of saturation of myoglobin with oxygen is plotted against oxygen pressure, a steady rise is observed until complete saturation is approached and the curve levels off. The oxygen-binding curve of myoglobin is thus said to be hyperbolic.

In contrast, the shape of the oxygen binding curve for hemoglobin is sigmoidal. This shape indicates that the binding of the first oxygen molecule facilitates the binding of the second oxygen, which facilitates the binding of the third, which in turn facilitates the binding of the fourth. This is precisely what is meant by the term cooperative binding. The function of hemoglobin is oxygen transport, and it must be able both to bind strongly to oxygen and to release oxygen easily, depending upon conditions. In the alveoli of lungs, the oxygen pressure is 100 torr.

At this pressure hemoglobin is 100% saturated with oxygen. In the capillaries of active muscles the pressure of oxygen is 20 torr, corresponding to less than 50% saturation of hemoglobin (which occurs at 26 torr). In other words, hemoglobin gives up oxygen easily in capillaries, where the need for oxygen is great. Structural changes during binding of small molecules are characteristic of allosteric proteins such as hemoglobin.

Hemoglobin has different quaternary structures in the bound (oxygenated) and unbound (deoxygenated) forms. The two β-chains are much closer to each other in oxygenated hemoglobin than in deoxygenated hemoglobin. The change is so marked that the two forms of hemoglobin have different crystal structures. Both H^+ and CO_2, which themselves bind to hemoglobin, affect the affinity of hemoglobin for oxygen by altering the protein's three-dimensional structure in subtle but important ways.

The effect of H^+ is called the Bohr effect after its discoverer, Christian Bohr. An increase in the concentration of H^+ reduces the oxygen affinity of hemoglobin. Actively metabolizing tissue, which requires oxygen, releases H'. Hemoglobin has a lower affinity for oxygen under these conditions and releases oxygen where it is needed.

Hemoglobin's acid-base properties affect, and are affected by its oxygen-binding properties. The oxygenated form of hemoglobin is a stronger acid (has a lower pK_a) than the deoxygenated form. In other words, deoxygenated hemoglobin has a higher affinity for H' than does the oxygenated form. Thus, changes in the quaternary structure of hemoglobin can modulate the buffering of blood through the hemoglobin molecule itself. The CO_2 produced by metabolism turn forms carbonic acid, H_2CO_3. The pK_a of H_2CO_3 is 6.35; the normal pH of blood is 7.4. As a result, about 90% of dissolved CO_2 will be present as the bicarbonate ion, HCO^-_3, releasing H^+.

The in vivo buffer system involving H_2CO_3 and HCO_3 in blood was discussed. The presence of larger amounts of H^+ as a result of CO_2 production favours the quaternary structure that is characteristic of deoxygenated hemoglobin. Hence, the affinity of hemoglobin for oxygen is lowered. The HCO_3^- is transported to the lungs, where it combines with H^+

released when hemoglobin is oxygenated, producing H_2CO_3. In turn, H_2CO_3 liberates CO_2, which is then exhaled. In the presence of large amounts of H^+ and CO_2, as in respiring tissue, hemoglobin releases oxygen.

The presence of large amounts of oxygen in the lungs reverses the process, causing hemoglobin to bind O_2. The oxygenated hemoglobin can then transport oxygen to the tissues. In blood, hemoglobin is bound to another ligand, 2,3-bisphosphoglycerate (BPG), with drastic effects on its oxygen-binding capacity. The binding is electrostatic; specific interactions take place between the negative charges on BPG and positive charges on the protein.

In the presence of BPG, the partial pressure at which 50% of hemoglobin is bound to oxygen is 26 torr. If BPG were not present in blood, the oxygen-binding capacity of hemoglobin would be much higher and little oxygen would be released in the capillaries. Stripped hemoglobin, which is isolated from blood and from which the endogenous BPG has been removed, displays this behaviour. BPG also plays a role in supplying a growing fetus with oxygen.

The fetus obtains oxygen from the mother's bloodstream by means of the placenta. Fetal hemoglobin (Hb F) has a higher affinity for oxygen than does maternal hemoglobin, allowing for efficient transfer of oxygen from the mother to the fetus.

Structure of proteins and contraction of muscles

Muscle cells are made up of fibers called myofibrils, which in turn are made up of repeating units called sarcomeres. In a sarcomere, thick and thin filaments lie parallel to each other, their ends partially overlapping. The thick filament consists of an aggregate of several molecules of the protein myosin. The thin filament consists of an aggregate of the protein actin. Other proteins that are involved in muscle contraction occur in the thin filament, but actin and myosin are the most important contributors to the process; we shall concentrate on them.

The presence of thick and thin filaments is responsible for the appearance of the sarcomere, characterized by two alternating bands of different degrees of darkness: a light I band and a heavy A band. The thick filament occurs by itself in the comparatively light H zone within the A band, and it is parallel to the thin filament where it overlaps it in the rest of the A band.

Within the H zone is the M line, another dark portion representing the over-lap of thick filaments. The I band consists of thin filaments only, but it is bisected by the dark Z line, an area rich in proteins other than actin. In the process of muscle contraction, the thick and thin filaments slide past one another. Their lengths do not change, but the amount of overlap increases as the muscle contracts. In the process, the mode of interaction of actin and myosin changes.

Myosin is principally a fibrous protein, but it has globular portions that appear as projections from the thick filament. The globular portions of myosin interact with actin to form an actomyosin complex. This process requires energy, provided by the hydrolysis of adenosine triphosphate (ATP) to adenosine diphosphate (ADP) and phosphate ion (Pi). When a muscle contracts, each globular region of myosin first tilts, then becomes detached from the site to which it was bound to the actin part of the actomyosin complex, and moves to a new site.

The tilting of the globular head of the myosin is considered to provide the power stroke for the contraction, a process that indirectly makes use of the energy obtained from ATP hydrolysis. Muscle contraction requires Ca^{2+} as well as energy. Calcium's role in muscle contraction involves the other protein components of the thin filament, troponin and tropomyosin.

An increase in the level of Ca^{2+} comes about as a result of the opening of Ca^{2+} channels in the sarcoplasmic reticulum, produced by the action of the neuromuscular junction. The high level of Ca^{2+} triggers a conformational change in troponin, a change that is passed along to tropomyosin. The conformational change in tropomyosin, in turn, triggers the tilting and sliding motion of myosin, which is the main feature of muscle contraction.

Protien synthesis

Transfer RNA (tRNA) molecules are specific for a given amino acid, are required to collect activated amino acids and deliver them, one at a time, to the ribosome, which itself consists of ribosomal RNA and protein. Here they are sequentially joined to synthesize the polypepticle chain of a protein. The sequence of amino acids, derived from the sequence of DNA bases, is specified by the genetic code, using the four RNA bases A, U, G, and C taken three at a time.

In the triplet code there are 64 possible "code words" called codons, of which 3 are stop signals and 61 specify the 20 amino acids with considerable redundancy. In the actual mechanism of translation, messenger RNA is temporarily bound to transfer RNA. The ribosome is the site of protein synthesis.

The mRNA and tRNA, which are bound to the ribosome by noncovalent interactions in the course of protein synthesis, are responsible for the correct order of amino acids in the growing protein chain. Before an amino acid can be incorporated into a growing protein chain, it must first be activated, a process involving both tRNA and a specific enzyme of the class known as aminoacyl-tRNA synthetases. The amino acid is covalently bonded to the tRNA in the process, forming an aminoacyl-tRNA.

The actual formation of the polypeptide chain occurs in three steps. In the initiation step, the first aminoacyl-tRNA is bound to the mRNA at the site that encodes the start of polypeptide synthesis. The mRNA and the ribosome are bound to each other. The next aminoacyl-tRNA forms a complex with the ribosome and with mRNA. The binding site for the second aminoacyl-tRNA is close to that for the first aminoacyl-tRNA. A pepticle bond is formed between the amino acids (this step is called chain elongation). The chain elongation process repeats itself until the polypeptide chain is complete. Finally, chain termination takes place. Each of these steps has many distinguishing features, and we shall look at each of them in detail.

Genetic code

The genetic message is contained in a triplet, nonoverlapping, commaless, degenerate universal code. Each of these terms has a definite meaning that describes the way in which the code is translated. A triplet code means that a sequence of three bases (called a codon) is needed to specify one amino acid. The term nonoverlapping indicates that no bases are shared between consecutive codons; the ribosome moves along the mRNA three bases at a time rather than one or two at a time.

If the ribosome moved along the mRNA more than three bases at a time, this situation would be referred to as "the presence of commas in the code." Since no intervening bases exist between codons, the code is commaless. In a degenerate code, more than one triplet can code for the same arnino acid. There are 64 (4 × 4 × 4) possible triplets of the four bases that occur in RNA, and all of them are used to code for 20 amino acids or for one of the three stop signals. A universal code is one that is the same in all organisms.

The universality of the code has been observed in viruses, prokaryotes, and eukaryotes; the only exceptions are the differences in a few codons seen in mitochondria. The evolutionary origin of these differences is not known at this writing. All 64 codons have been assigned meanings, with 61 of them coding for amino acids and the remaining 3 serving as the termination signals.

Two amino acids, tryptophan and methionine, have only one codon each, but the rest have more than one. A single amino acid can have as many as six codons, as is the case with leucine, serine, and arginine. Multiple codons for a single amino acid are not randomly distributed in Table 1 but tend to have one or two bases in common. The bases that are common to several codons are usually the first and second bases, with more room for variation in the third base, which is called the "wobble" base. The assignment of triplets in the genetic code is based on several types of' experiment. One of the most significant involves the use of synthetic polyribonucleotides as messengers. This approach can give some information about the nature of the code, but ambiguities remain.

Unambiguous assignments require another important type of experiment that is based on binding studies involving tRNAs and triplets synthesized in the laboratory. When homopolynucleotides (polyribonucleoticles that contain only one type of base) are used as a synthetic mRNA in laboratory systems for polypepticle synthesis, homopolypeptides (polypeptides that contain only one kind of amino acid) are produced. When poly A is the messenger, the product is polyphenylalanine.

With poly A as the messenger, polylysine is formed. The product for poly C is polyproline; with poly G solubility problems interfere, but it has been determined that a GGG triplet codes for glycine. When an alternating copolymer (a polymer with an alternating sequence of two bases) is the messenger, the product is an alternating polypeptide (a polypeptide with an alternating sequence of two amino acids).

For example, when the sequence of the polynucleoticle is -ACACACACACACACACACACAC-, the polypepticle produced is poly (thr-his), with the alternating sequence threonine-histidine. There are two types of coding triplets in this polynucleotide, ACA and CAC, but this experiment cannot establish which one codes for threonine and which one for histidine. More information is needed for an unambiguous assignment, but it is interesting that this result proves that the code is a triplet code.

If it were a doublet code, the product would be a mixture of two homopolymers, on specified by the codon AC and the other by the codon CA. Other methods are needed to answer the remaining questions about codon assignment. One of the most useful methods is the binding assay. This technique depends on the fact that aminoacyl-tRNAs bind strongly to ribosomes in the presence of trinucleotides. In this situation, the trinucleotide plays the role of an mRNA codon.

The possible trinucleotides are synthesized by chemical methods, and binding assays are carried out with each type of trinucleotide. Aminoacyl-tRNAs are tested for their ability to bind in the presence of a given trinucleotide. For example, if the aminoacyl tRNA for isoleucine binds to the ribosome in the presence of the trinucleotide AUC, the sequence AUC is established as a codon for isoleucine. About 50 of the 64 codons have been identified by this method.

Some tRNAs bond to one codon exclusively, but many of them can recognize more than one codon because of variations in the allowed pattern of hydrogen bonding. This variation is called "wobble", and it applies to the first base of an anticodon, the one at the 5' end, but not to the second or the third base. Recall that mRNA is read from the 5' to the 3' end. The first (wobble) base of the anticodon hydrogen-bonds to the third base of the codon, the one at the 3' end. The base in the wobble position of the anticodon can base pair with several different bases in the codon, not just the base specified by Watson-Crick base pairing.

When the wobble position of the anticodon is occupied by uridine, it can base-pair not only with adenosine, as expected, but also with guanosine, the other purine. When the wobble position is occupied by guanosine, it can base-pair with cytidine, as expected, and also with uridine, the other pyrimidine.

The purine nucleoside inosine frequently occurs in the wobble position in many tRNAs, and it can base-pair with adenosine, cytidine, and uridine in the codon. Adenosine and cytidine do not form any base pairs other than the expected ones with uridine and guanosine, respectively.

To summarize, when the wobble position is occupied by 1, G, or U, variations in hydrogen bonding are allowed; when the wobble position is occupied by A or C, these variations do not occur. The wobble hypothesis provides insight into some aspects of the degeneracy of the code.

In many, but not all, cases the degenerate codons for a given amino acid differ in the third base, the one that pairs with the wobble base of the anticodon. Fewer different tRNAs are needed, since a given tRNA can base -pair with several codons. As a result, a cell would have to invest less energy in the synthesis of needed tRNAs.

The existence of wobble also minimizes the damage that can be caused by misreading of the code. If, for example, a leucine codon, CUU, were to be misread as CUC, CUA, or CUG during transcription of mRNA, this codon would still be translated as leucine during protein synthesis; no damage to the organism would occur.

The activation of the amino acid and the formation of the aminoacyl-tRNA take place in two separate steps, both of which are catalyzed by aminoacyl-tRNA synthetase.

First, the amino acid forms a covalent bond to an adenine nucleotide, producing an aminoacyl-AMP. The free energy of hydrolysis of ATP provides energy for bond formation. The aminoacyl moiety is then transferred to tRNA, forming an aminoacyl-tRNA.

$$\text{Amino acid} + \text{ATP} \rightarrow \text{aminoacyl-AMP} + \text{PPi}$$
$$\text{Aminoacyl-AMP} + \text{tRNA} \rightarrow \text{aminoacyl-tRNA} + \text{AMP}$$

$$\text{Amino acid} + \text{ATP} + \text{tRNA} \rightarrow \text{aminoacyl-tRNA} + \text{AMP} + \text{PP}_i$$

Aminoacyl-AMP is a mixed anhydride of a carboxylic and a phosporic acid.

Since anhydrides are reactive compounds, the free energy change for the hydrolysis of aminoacyl-AMP favours the second step of the overall reaction. Another point that favours the process is the energy released when pyrophosphate (PP_i) is hydrolyzed to orthophosphate (P_i) to replenish the phosphate pool in the cell.

The hydrolysis of pyrophosphate removes one of the products of the reaction and thus favours the forward reaction rather than the reverse reaction. The synthetase enzyme requires Mg^{2+} and is highly specific both for the amino acid and for the tRNA. A separate synthetase exists for each amino acid.

In the second part of the reaction, an ester linkage is formed between the amino acid and the 3'-hydroxyl end of the tRNA. Several tRNAs can exist for each amino acid, but a given tRNA will not bond to more than one amino acid. Each synthetase has a high degree of specificity for the correct tRNA and the correct amino acid. The specificity of the enzyme contributes to the accuracy of the translation process. This extra level of proofreading is part of what is sometimes called the "Second Genetic Code".

The two-stage reaction allows for selectivity to operate at two levels: that of the amino acid and that of the tRNA. The specificity of the first stage makes use of the fact that the aminoacyl-AMP remains bound to the enzyme. For example, isoleucyl-tRNA synthethase can form a bond between AMP and isoleucine or the structurally similar valine. The tRNA synthetase has two active sites, one for acylation of AMP and one for hydrolysis of the aminoacyl-AMP. Valyl-AMP is selectively hydrolyzed at the hydrolytic site rather than isoleucyl-AMP. The second aspect of selectivity depends on the selective recognition of tRNAs by aminoacyl-tRNA synthetases.

There are specific binding sites on tRNAs that are recognized by aminoacyl-tRNA synthetases. The exact position of the recognition site varies with different synthetases, with the role of the anticodon in binding of tRNA to the synthetase having the greatest variability. Note that there is a high energy cost, paid in the form of

ATP hydrolysis, needed to achieve this degree of fidelity in protein synthesis. The payoff for the cell is that energy is not wasted in producing defective proteins.

We shall use *Escherichia coli* as our principal example, since all aspects of protein synthesis have been most extensively studied in this bacterium. In all organisms the synthesis of polypeptide chains starts at the *N*-terminal end; the chain grows from the N-terminal to the C-terminal end.

In prokaryotes the initial N-terminal amino acid of all proteins is *N*-formylmethionine (fmet). However, this residue can be, and often is, removed by posttranslational processing after the polypeptide chain is synthesized. There are two different tRNAs for methionine in *E. coli*, one for unmodified methionine and one for N-formylmethionine. These two tRNAs are called $tRNA^{met}$ and $tRNA^{fmet}$, respectively (the superscript identifies the tRNA). The aminoacyl-tRNAs that they form with methionine are called met-$tRNA^{met}$ and met-$tRNA^{fmet}$, respectively (the prefix identifies the bound amino acid).

In the case of met-$tRNA^{fmet}$ a formylation reaction takes place after methionine is bonded to the tRNA, producing *N*-formylmethionine-$tRNA^{fmet}$ (fmet-$tRNA^{fmet}$). The source of the formyl group is N^{10}-formyltetrahydrofolate. Methionine bound to $tRNA^{met}$ is not formylated. Both tRNAs ($tRNA^{met}$ and $tRNA^{fmet}$) contain a specific sequence of three bases (a triplet), 3′-UAC-5', which base-pairs with the sequence 5'-AUG-3' in the mRNA sequence. The $tRNA^{fmet}$ triplet in question, UAC, recognizes the AUG triplet, which is the start signal when it occurs at the beginning of the mRNA sequence that directs the synthesis of the polypeptide.

The same UAC triplet in $tRNA^{met}$ recognizes the AUG triplet when it is found in an internal position in the mRNA sequence. The start signal is preceded by a leader segment of mRNA, the Shine-Dalgarno sequence (5'GGAGGU-3'), which usually lies about ten nucleotides upstream of the AUG start signal (nearer the 5' end of the RNA). The genetic message of mRNA is read from the 5' to the 3' end. A portion of the mRNA leader segment binds to the 30S subunit of the ribosome by forming base pairs with the 3' portion of the 16S rRNA of the subunit.

The mRNA triplet, the codon, specifies the nature of the amino acid to be added to the growing polypepticle chain. The tRNA triplet that hydrogenbonds to the codon is called the anticodon. Note that the anticodon is antiparallel and complementary to the mRNA codon. The start of polypeptide synthesis requires formation of an initiation complex.

At least eight components enter into the formation of the initiation complex, including mRNA, the 30S ribosomal subunit, fmet-$tRNA^{fmet}$ GTP, and three protein initiation factors, called IF-I, IF-2, and IF-3. The IF-3 protein facilitates the binding of mRNA to the 30S ribosomal subunit. It also appears to prevent premature binding of the. 50S subunit, which takes place in a subsequent step of the initiation process. IF-2 binds GTP and aids in the selection of the initiator tRNA fmet-$tRNA^{fmet}$") from all the other aminoacylated tRNAs available.

The function of IF-I is less clear; it appears to bind to IF-3 and to IF-2 and facilitates the action of both. The resulting combination of mRNA, the 30S ribosomal subunit, and fmet-$tRNA^{fmet}$ is the 30S initiation complex. A 50S ribosomal subunit binds to the 30S initiation complex to produce the 70S initiation complex. The hydrolysis of GTP to GDP and P_i favours the process by providing energy; the initiation factors are released at the same time.

The process of chain initiation in eukaryotes has not been studied to the extent it has been in bacteria, but several differences are known to exist. Methionine, not *N*-formylmethionine, is the amino acid used for initiation in eukaryotes. Two tRNAs for methionine are found in eukaryotes, and with one of them, the formylase enzyme from *E. coli* can produce fmet-tRNA. Apparently eukaryotes have lost the ability to carry out the reaction because of the lack of the proper enzyme rather than because of the properties of the tRNA.

Eukaryotes apparently use a "scanning" mechanism to locate the initiator AUG codon. The 40S ribosomal subunit (recall the size differences between prokaryotic and eukaryotic ribosomes and their subunits) moves down the message from the capped 5'-end until it encounters an AUG with the appropriate base sequence surrounding it. This is often, though not always, the first AUG. After this AUG is found, the 60S subunit and protein factors bind so that initiation can begin. There is no Shine-Dalgarno sequence in eukaryotes.

There are at least eight initiation factors in eukaryotes; they are designated eIF-1, eIF-2, and so on, with the "e" (for eukaryotic) used to distinguish them from prokaryotic initiation factors.

The elongation phase of prokaryotic protein synthesis makes use of the fact that three binding sites for tRNA are present on the 50S subunit of the 70S ribosome. The three tRNA binding sites are called the P (peptidyl) site, the A (aminoacyl) site, and the E (exit) site. The P site binds at tRNA that carries a peptide chain, and the A site binds an aminoacyl-tRNA. The *E* site carries an uncharged tRNA that is about to be released from the ribosome.

Chain elongation begins with the addition of the second amino acid specified by the mRNA to the 70S initiation complex. The *P* site on the ribosome is the one initially occupied by the fmet-tRNA^{fmet} in the 70S initiation complex. The second aminoacyl-tRNA binds at the A site. A triplet of tRNA bases (the anticodon AAA in our example) forms hydrogen bonds with a triplet of mRNA bases (UUU, the codon for phenylalanine, in this example).

In addition, GTP and two protein elongation factors, EF-Tu and EF-Ts (temperature-unstable and temperature-stable elongation factors, respectively), are required for the binding of aminoacyl-tRNAs to the A site. GTP is hydrolyzed to GDP and P_i in the process. A *peptide bond* is formed in a reaction catalyzed by peptidyl transferase, which is a part of the 50S subunit.

Recent experiments indicate that the peptidyl transferase is the 23S rRNA. This result is a striking example of the catalytic activity of rRNA. The carboxyl group of the *N*-formylmethionyl residue forms a bond with the amino group of the amino acid bound to the tRNA at the A site. There are now a dipeptidyl-tRNA at the A site and a tRNA with no amino acid attached (an uncharged tRNA) at the P site. A translocation step then takes place before another amino acid can be added to the growing chain.

In the process the uncharged tRNA moves from the P site to the E site, from which it is subsequently released; the peptidyl-tRNA moves from the A site to the vacated P site. In addition, the mRNA moves with respect to the ribosome. Another elongation factor, EF-G, also a protein, is required at this point, and once again GTP is hydrolyzed to GDP and P_i.

The three steps of the chain elongation process are aminoacyl-tRNA binding, peptide bond formation, and translocation. They are repeated for each amino acid specified by the genetic message of the mRNA until the stop signal is reached. Puromycin is a structural analogue of the 3' end of an aminoacyl-tRNA, making it a useful probe to study chain elongation. In an experiment of this sort, puromycin binds to the A site, and a peptide bond is formed between the C-terminal of the growing polypeptide and the puromycin.

The peptidyl puromycin is weakly bound to the ribosome and dissociates from it easily, resulting in premature termination and a defective protein. Puromycin also binds to the P site and blocks the translocation process, although it does not react with peptidyl-tRNA in this case. The existence of A and P sites was determined by these experiments with puromycin.

The main features of the elongation process are the same in prokaryotes and eukaryotes, but the details differ. These differences can be seen in the response to inhibitors of protein synthesis and to toxins. The antibiotic chloramphenicol (a trade name is Chloromycetin) binds to the A site and inhibits peptidyl transferase activity in prokaryotes, but not in eukaryotes. This property has made chloramphenicol useful in treating bacterial infections.

One of two protein release factors (RF-1 or RF-2) is required, as is GTP, which is bound to a third release factor, RF-3. RF-1 binds to UAA and UAG, and RF-2 binds to UAA and UGA. RF-3 does not bind to any codon, but it does facilitate the activity of the other two release factors. Either RF-1 or RF-2 is bound near the A site of the ribosome when one of the termination codons is reached. The release factor not only blocks the binding of a new aminoacyl-tRNA but also affects the activity of the peptidyl transferase, so that the bond between the carboxyl end of the peptide and the tRNA is hydrolyzed. GTP is hydrolyzed in the process.